The Enterprise of Flight

SMITHSONIAN HISTORY OF AVIATION AND SPACEFLIGHT SERIES

Dominick A. Pisano and Allan A. Needell, Series Editors

Since the Wright brothers' first flight, air and space technologies have been central in creating the modern world. Aviation and spaceflight have transformed our lives—our conceptions of time and distance, our daily routines, and the conduct of exploration, business, and war. The Smithsonian History of Aviation and Spaceflight Series publishes substantive works that further our understanding of these transformations in their social, cultural, political, and military contexts.

THE ENTERPRISE OF FLIGHT
THE AMERICAN AVIATION AND AEROSPACE INDUSTRY

ROGER E. BILSTEIN

SMITHSONIAN INSTITUTION PRESS • WASHINGTON AND LONDON

Especially for Linda

© 2001 Roger E. Bilstein
All rights reserved

Library of Congress Cataloging-in-Publication Data
Bilstein, Roger E.
   The enterprise of flight : the American aviation and aerospace industry / Roger E. Bilstein.
     p. cm.
   Previous ed. published under: American aerospace industry. 1996.
   Includes bibliographical references and index.
   ISBN 1-56098-964-5 (alk. paper)
     1. Aerospace industries—United States—History.   2. Aircraft industry—United States—
     History.   I. Bilstein, Roger E. American aerospace industry.   II. Title.
   HD9711.5.U6 B55   2001
   338.4.′76291′0973—dc21   2001031082

British Library Cataloguing-in-Publication Data available

Manufactured in the United States of America
08 07 06 05 04 03 02 01   5 4 3 2 1

∞ The paper used in this publication meets the minimum requirements of the American National Standard for Information Sciences—Permanence of Paper for Printed Library Materials ANSI A39.48-1984.

For permission to reproduce illustrations appearing in this book, please correspond directly with the owners of the works, as listed in the individual captions. The Smithsonian Institution Press does not retain reproduction rights for these illustrations individually or maintain a file of addresses for photo sources.

CONTENTS

Acknowledgments	vii
Introduction	x
1. The Flamboyant Years, 1900–1919	1
2. Outlines of an Industry, 1919–1933	21
3. Aviation and Airpower, 1933–1945	49
4. Cold War Responses, 1945–1969	79
5. Rockets and Space, 1926–1969	108
6. Airliners and Light Planes, 1945–1969	135
7. Changing Horizons, 1970s/1980s	160
8. Global Perspectives, 1980s/1990s	187
9. Twenty-First-Century Transitions	209
Appendix	225
Chronology	234
Notes	240
Selected Bibliography	271
Index	275

ACKNOWLEDGMENTS

WHEN THIS BOOK BEGAN, Twayne's series on the Evolution of American Business was under the direction of Professor Edwin J. Perkins, of the University of Southern California, who provided wise suggestions for its general outline. He was succeeded by Professor Kenneth Lipartito, University of Houston, who offered much appreciated encouragement and editorial guidance as the manuscript proceeded. I have often profited from conversations and correspondence with aviation historian Dr. Richard K. Smith, even though he probably would not agree with all that I have written. In a remarkable act of generosity, Dr. William Leary, University of Georgia, volunteered to read and comment on the entire manuscript.

My own approach to this project has been inevitably shaped by previous books and articles that I have written. Indeed, in the preparation of the index for *Flight in America: From the Wrights to the Astronauts*, 1st ed. (Baltimore: The Johns Hopkins University Press, 1984), I realized that a pervasive theme of significant European influences on American aviation had evolved in that book with no overt intent on my part. I continued to explore this aspect of the American aviation experience in subsequent books and articles, and the European thread appears as a strong subtheme in the present study. These subsequent books and articles are cited in the backnotes and I wish to acknowledge the relevant editors for permission to incorporate certain information and observations that appear in this publication. My understanding of the European heritage was enhanced by an interview with Air Commodore A.D.A. Honley, Society of British Aerospace Constructors (London, 1988); by discussions with research specialists at the Royal Air Force Museum (Hendon, 1991); and by conversations with corporate historians at Rolls-Royce (Derby, 1991), especially James Cownie and David Birch. I also explored this topic and other areas of aviation history in many lively conversations with Dr. Richard P. Hallion, Air Force Historian, U.S. Air Force.

During the summer of 1989, a grant from the Faculty Research Support Fund of the University of Houston–Clear Lake allowed time to develop bibliographical resources and research notes for the overall study. Jay Miller (Dallas) again opened his home and his remarkable archival collection during more than one visit. In 1991, during the spring semester, I was able to make progress on the manuscript as the recipient of the American Historical Association's Senior Fellowship in Aerospace History. During the academic year 1992–93, I received a Faculty Development Leave from my university and was also named to the Charles A. Lindbergh Chair in Aerospace History at the National Air and Space Museum, Smithsonian Institution. To my colleagues at the National Air and Space Museum, I owe a great deal. Dr. Tom Crouch, chair of the Department of Aeronautics, skillfully arranged the details of my appointment and generously shared insights about aviation history. Dr. Dominick Pisano exchanged opinions with me about social and cultural themes, and Dr. Peter Jakab pointed me toward materials on early flight. Also, I wish to acknowledge the wisdom of Robert van der Linden on early airline technology; R.E.G. Davies on international airliner technology; Dorothy Cochrane on general aviation trends; Dr. Von Hardesty on air and space topics; Rick Leyes on power plants; Russ Lee on industrial propaganda; Dr. Mike Neufeld on military topics and European sources; Alex Spencer on helicopters and other subjects; Tom Dietz on scale models; and Joanne Gernstein on aviation publicity. Frank Winter, of the museum's Department of Space History, proved an invaluable source of assistance about early industrial developments in rocketry. Dr. Claudio Segre (University of Texas–Austin) and Dr. John Anderson (University of Maryland), who held continuing research appointments at the museum, graciously exchanged observations with me about a fascinating variety of aeronautical topics. Mary Pavlovich (NASM Library) and Dan Hagedorn (NASM Archives) helped to locate esoteric sources. Also at the museum, Mildred Anita Mason, Mary Nettleton, and Collette Williams typed sections of a preliminary manuscript.

At the Neumann Library, University of Houston–Clear Lake, reference librarians and interlibrary loan librarians invariably helped me to track down obscure bits of information and book titles. For their assistance and interest in my project, I particularly thank Gay Carter and Patricia Pate, reference librarians. Over the years a number of students have shared in interests in aviation history; for their contributions to my overall understanding of the subject, I thank Cecil Arnold, Edward Morison, and Byron Myers. Dr. Shirley Paolini (Dean, School of Humanities and Human Sciences), and Dr. Carol Snyder (Associate Dean of Academic Affairs, HSH), graciously supported applications and other arrangements for grants to provide time for research and writing. Anne

Seaman (Business Coordinator, HSH) helped arrange typing support at a crucial time. The final draft was typed by Diane DeVusser and Margo Mosley; it is to Margo that I owe a special debt of gratitude for reconstructing the text from scattered files in a word processor and for completing the final version of the typescript.

During the academic year 1995–96, I served as Visiting Professor of History at the United States Air Force Air War College, Maxwell AFB, Alabama. Dr. Alexander Cochran, chair of the Department of Strategy, Doctrine, and Air Power, diplomatically made allowances for the time necessary to make final editorial changes to the manuscript; Dr. James Mowbray and Colonel Robin Read, USAF, cheerfully responded to frequent questions about aviation technology and trivia.

A variety of manufacturers responded with photographs and information. I would like to thank them collectively, and direct the attention of readers to the credit lines that accompany illustrations used in the text. Personnel at the Aerospace Industries Association repeatedly responded to requests for miscellaneous data. At Twayne Publishers, Carol Chin, as senior editor, deserves credit for empathy and tolerance when unforseen circumstances created a long delay in the manuscript's progress; her successor, Margaret Dornfeld, proved just as understanding and supportive. Nan Gatewood, the copy editor, saved me from many embarrassing problems.

My greatest thanks go to Linda, who shares my life, shares my fascination with the world of flight, and who remains patient about books and notes scattered about the house. I also want to acknowledge our children, Paula and Alex, who still listen to repetitious aeronautical anecdotes.

Finally, any errors of fact or interpretation remain mine alone.

Roger E. Bilstein
Montgomery, Alabama, 1996

Note on the Second Edition

It has been a special pleasure to develop this new and retitled edition of *The American Aerospace Industry* for the Smithsonian Institution Press. My thanks to all those acknowledged in the first edition still stand. Gerald Churchill, reference librarian at the University of Houston-Clear Lake, helped track down new information. For the revised edition, I owe a special debt to Mark Gatlin at the Smithsonian Institution Press for his unfailing humor and hard work in shepherding the book over various editorial hurdles.

Roger E. Bilstein
Dripping Springs, Texas, 2001

Introduction

DURING THE MID-1990S, news stories about the aviation and aerospace business in the United States often included commentary on two themes: mergers and internationalization of the industry. By the turn of the century, these topics continued to be significant news items, along with a rash of new headlines about stiffer competition from overseas, especially in the case of European airliners produced by Airbus that challenged the dominating position long held by American manufacturers like Boeing. Controversies about costly military aircraft continued, and speculation about the role of American equipment competing with European designs to equip foreign air forces sharpened debates about military strategy and overseas trade. In the general aviation sector, a strong market for executive jets augmented production figures and balance sheets. The exploration of space experienced both successes and failures, although the operational debut of the International Space Station, which united astronauts and technologies of former Cold War protagonists, represented one of the most symbolic events of the aerospace industry as the twentieth century gave way to the new millennium.[1]

Technological Challenges: Domestic and Foreign

In the production of large airliners, the reality of the European Airbus challenge dramatically hit home in 1996 when the historic American company McDonnell Douglas suddenly announced a merger with its archrival, Boeing. Even though McDonnell Douglas had experienced prior economic strains, the company exhibited a renewed vigor in the mid-1990s, with continuing sales of missiles, airliners, fighters, a new Navy trainer (the T-45, based on British technology), and a new transport (the C-17). Underneath the surface, however, long-term problems with debts, accounting practices, management turmoil, and languid technological development finally caught up with the company's leadership. Boeing bought out its perennial competitor for $13 billion in a deal that became official during 1997. Other significant mergers followed, such as General Electric's acquisition of Honeywell, although none possessed the dramatic dimensions of Boeing's coup. Surveying a landscape increasingly devoid of so many familiar corporate names, analysts also noted that a similar phenomenon had been occurring among second- and third-tier suppliers as well.

In Europe a combination of economic factors and alarm over the formidable competition posed by giant American aerospace companies led to similar strategies of acquisitions and mergers. After buying out several firms, British Aerospace reformed as BAE Systems in January 1999. Its American acquisitions made it a significant contractor to the U.S. armed services. Across the English Channel, other European aerospace firms fretted about potential British dominance in their own backyards. Within months, Germany, France, and Spain organized a counterpart, the European Aeronautics, Defense, and Space Company (EADS). At the same time, both BAE and EADS cooperated in complex preparations for a strengthened Airbus Industrie consortium and announced plans for a superjumbo to dethrone the 747 and to challenge Boeing's lofty position as the builder of the world's only jumbo jet.

Although many circumstances led to the dramatic demise of McDonnell Douglas, the long shadow of competition from Airbus clearly represented a major factor. Following its formation as a consortium of European manufacturers in 1970, Airbus shrewdly evolved a series of airliners that shared the same basic cockpit configuration and electronic control systems. Development of a superjumbo became a major goal. The Europeans wanted this plane, initially called the Airbus 3XX, to eclipse the 747 in size, range, and global prestige. By the end of 2000, Airbus had collected enough orders (fifty) to justify an official commitment to the superjumbo, now designated the Airbus A380.

The surge in orders for all of Airbus's planes, including the A380, frustrated not only Boeing executives but also many union leaders, who felt that job cuts on Boeing's commercial production lines could be traced to unfair competition. According to American critics, Airbus not only received unfair state subsidies but also functioned within a complex economic scheme that gave the company more leeway to promise huge discounts to overseas clients. Even though the A380 program might involve 22,000 jobs in the United States, the controversy led U.S. trade representatives to file briefs at the World Trade Organization asking for clarification and details of Airbus's financing. As the issue continued to smolder, Airbus and Boeing hewed to divergent strategies in the high stakes game of selling airliners. Airbus contended that rising numbers of travelers between major hubs dictated huge airliners like the 600-passenger A380. Boeing's studies indicated a trend for direct flights to smaller cities, dictating a market for more planes with fewer seats. Each manufacturer made a gamble representing billions of dollars and global prestige on the outcome of their decision.[2]

The Military Market Place

The U.S. Air Force continued to wrestle with tight budgets, shifting doctrinal concepts, and plans for aircraft presumably required in the twenty-first century. A top priority was a new air superiority fighter. Development of the F-22 Raptor, manufactured by Lockheed with Boeing as a major partner, moved ahead as several pre-production aircraft proceeded to wring out electronic systems and to establish operational data. With its "stealthy" design, thrust-vectoring system, and advanced avionics suite, the Raptor's highly sophisticated design for multiple missions came with an equally high price of $85 million to $100 million per plane, depending on how many the U.S. Air Force could fund. The Raptor's price tag continued to trigger budgetary heartburn in congressional committee hearings and to raise questions about funding for another new program for a combat aircraft, the Joint Strike Fighter (JSF)>.

Specifications for the JSF, which was a unique design effort, answered requirements for the U.S. Air Force, Navy, and Marine Corps: a stealthy, low-level attack plane that could hold its own in combat with enemy aircraft. The version for the U.S. Marine Corps included short takeoff and vertical landing capability (STOVL). Because this latter characteristic also met requirements for the Royal Air Force, the British played a key role in supplying STOVL technology, financing, and design features, making the JSF a four-way international project. Boeing and Lockheed built competing prototypes— the X-32 and the X-35, respectively—that began test flights early in 2001. Economists estimated the cost of production versions at $28 million to $35 million per plane.

Other factors impacted the development of both the F-22 and the JSF. Partisans of the F-22 argued for acceptance and production as soon as possible in 2001, with operational squadrons scheduled for service by 2005–06. Otherwise, frontline U.S. Air Force fighters like the F-15 would be a quarter of a century old and severely disadvantaged if facing export versions of the newer generation of foreign combat planes such as the British/German/Italian Eurofighter, the French Rafale, or the Swedish Gripen. At the same time, less expensive planes like the JSF could reasonably satisfy the tactical requirements of various foreign air forces. In the twenty-first century, American aerospace pundits thought that such a plane was essential if the

United States wanted to maintain its role as a principal supplier of combat planes to friendly nations and to promote technical prowess in an industry known for generating significant spin-off technologies.

New technologies frequently carried the risk of trial and tribulation. In the case of the twin-engine Bell/Boeing V-22 Osprey tilt-rotor aircraft, tragic events included the fatal crashes of two early operational vehicles during 2000. With the U.S. Marine Corps pinning its hopes on the Osprey as a replacement for its aging troop helicopters, service experts and corporate engineers worked overtime to pinpoint and solve its problems. All three armed services in the United States increased their work in the field of uninhabited air vehicles (UAV) directed by ground-based controllers. Boeing, Lockheed, and Northrop Grumman all had new UAV designs in progress for vehicles with wingspans from a few dozen feet up to more than 100 feet.[3]

The General Aviation Business

Even though production of single-engine light planes lagged until passage in 1994 of the General Aviation Revitalization Act, which established limited liability protection for manufacturers, general aviation operations throughout the country convincingly demonstrated the value of those planes in modern life. Activities ranged from crop treatment and air ambulance service to forest fire control. Still, the majority of the general aviation fleet contributed to the rising numbers of hours spent in business flying. Clearly, the 1994 legislation stimulated expectations for builders of piston-engine aircraft such as Mooney, Piper, Beechcraft, and Cessna. Specialized designs from Maule, who built rugged planes for bush pilots, and from Aviat Aircraft, suppliers of bush planes and aerobatic planes, also anticipated stronger sales. Similarly, manufacturers of high-performance, single-engine planes such as Bellanca and Commander Aircraft expanded their marketing campaigns. A few new entrants, such as Cirrus Design Corporation, manufacturers of a plane featuring an integral parachute that deployed in emergencies, and Lancair, who delivered complete airplanes as well as supplying build-it-youself kits, found additional buyers.

The demand for expensive business jets became positively bullish by 2000, with sales in the $8 billion range—about triple the annual sales in the years before 1996. The strength of the demand led Boeing to offer an executive version of its 737 airliner, with options for amenities like king-size beds, showers, and other opulent touches. But the majority of sales involved smaller but still elegantly appointed aircraft for corporate officers. Several factors accounted for the sales of hundreds of new executive planes. First, many models of early jets delivered in the 1960s finally were beginning to reach the end of their useful lives. Second, reductions in regional airline service led numerous corporations to depend on their own executive aircraft. Builders like Raytheon/Beech, Bombardier, Cessna, Gulfstream, and others offered a wide array of models to suit a company's particular travel requirements and pocketbook. Rising demand for such planes was one of the factors that prompted General Dynamics to reenter the plane business by purchasing Gulfstream in 1999 for $5.3 billion, more than twice the company's sales during the prior year.

As for the world of helicopters, analysts noted a stable if not spectacular market during the late 1990s, with ongoing military deliveries. National prosperity during the 1990s sustained sales of civilian machines for executive transportation. Additionally, the American market recorded a rising demand for helicopters in law enforcement. Also, the closure of many rural hospitals, plus the consolidation of medical centers in urban areas, contributed to reliance on helicopters for ambulance transfer and emergency operations. Consequently, manufacturers made five hundred deliveries in 1998 and 1999 and reported total sales of about $1.3 billion each year.[4]

Rockets, Missiles, and Space

As the confrontation between the former Soviet Union and the United States became less intense, the launch rate for military satellites began to trail off, although it by no means disappeared. Military satellites for communications, early warning, and navigation possessed design lifetimes of four to ten years. Aging equipment, plus new technologies, buttressed an ongoing round of systems development, fabrication, and launches. The proliferation of commercial communications satellites generated a steady business for builders of launch vehicles. Large, established companies like Boeing and Lockheed Martin continued to dominate launch services in the United States, although smaller companies like Orbital Sciences Corporation were able to establish a presence in this highly competitive sector. A number of joint ventures with Russians and other foreign partners dramatized Cold War changes. Boeing had its own array of international schemes, including a 40 percent stake in the unique SeaLaunch program, a system to launch payloads from a floating platform positioned near the equator, about 1,400 miles southeast of Hawaii. The site imparted extra momentum to launch vehicles due to the earth's rotation, thus allowing engineers to plan heavier payloads or higher orbits. Boeing held overall responsibility for management and integration, including manufacture of numerous components. A Norwegian firm, Kvaener A/S, modified an offshore oil drilling rig as the launch center and outfitted a command vessel to transport the launch vehicle segments. Russian and Ukrainian firms supplied the upper stages.

Meanwhile, the push for peaceful uses of space spawned a panoply of civilian satellites that spun myriad orbits above the earth. From 1997 to 1999, about 135 satellites per year arrived in space; the annual value of these payloads ranged from $10 billion to $11 billion during the late 1990s and the first part of the twenty-first century. Commercial satellites provided a broad range of services: telecommunications, broadcasting, internet access, and imaging.

In an era of microscopic analysis of budgets, NASA undertook a major institutional change by gathering numerous support contracts under one roof and surrendering some of its own day-to-day operational functions to a single entity. The United Space Alliance (USA) was organized in 1996 in Houston by Lockheed Martin and Boeing to take over management of the space shuttle through a ten-year, $12 billion contract. NASA continued to launch a series of space exploration junkets, sending various probes to the Sun, Saturn, and other targets such as Mars. Although some of the Mars missions represented triumphs, expensive failures during 1999 (traced to confusion in computer programming) prompted NASA to overhaul its management approach of "faster, better, cheaper."

Against this background, NASA's adventures with the International Space Station (ISS) came under closer congressional scrutiny. Although Russia continued to be a key partner, providing launch services for resupply of the ISS in orbit and fabrication of the first crew modules, Russian bureaucracy and sporadic budgetary emergencies kept critical hardware lagging by as much as two years. In the meantime, between 1995 and 1999, seven American astronauts spent time aboard the aging Russian space station, *Mir*. Despite a rash of problems aboard the crotchety *Mir*, these excursions helped iron out many operational protocols and hardware issues involving Russian and American crews for future ISS missions. By the end of 2000, all the requisite modules finally reached orbit, carried by either the space shuttle or expendable Russian boosters. During November the first three-person crew (two Russians and one American) arrived aboard a Russian spacecraft to occupy the ISS for its formal inauguration as an international outpost, keeping station some 235 miles above the earth. With its solar wings (38 feet wide and 240 feet from tip to tip), the ISS became one of the brightest objects to be seen in the night sky.

But the apparent success of the ISS and other NASA space probes constituted a significant aspect of growing controversy about the role of NASA and the aerospace community relative to the decline of aeronautical research. With the demise of the Cold War, a decline in defense-related aeronautical research—from $641 million in 1996 to about $400 million in

2000—inevitably followed. However, NASA's efforts in this field also declined by about one-third as well, from $1 billion per year in the early 1990s to $640 million in 2000. During the 1990s, the number of bachelor and master's degrees awarded in aerospace disciplines also experienced a precipitous slide. By way of contrast, editors of *Aviation Week and Space Technology* warned that Europe's leadership had targeted the aerospace industry for special attention and increased funding for research and development. Furthermore, the editors declared, it was not surprising that America's share of the global aerospace market—more than 70 percent in the 1980s—had slipped to just a bit more than 50 percent. The aerospace industry, the editors concluded, had reached a crisis. Leading professional societies in America also hoisted warning flags.[5]

Critiques of America's declining status in comparison with the momentum achieved across the Atlantic held a certain irony. In the first two decades of the twentieth century, similar complaints had prompted the United States to assess European progress with care and to consider ways and means of improving America's position. One of the most influential developments involved the formation in 1915 of the National Advisory Committee for Aeronautics, the forerunner of NASA. The chapters that follow summarize the saga of America's enterprise in flight, a venture that encountered significant European influences from its beginning.

Notes

1. In preparing this overview, I frequently turned to *Aerospace Source Book 2000,* the latest edition of an annual, *Aerospace Source Book,* published by the well-known periodical *Aviation Week and Space Technology* (hereafter cited as *AWST*). In addition to succinct reviews of the aviation and aerospace industry's major business segments, these yearly compendiums include invaluable lists of current products and sketches of principal manufacturers as well as airline operators. I have also relied on an e-mail source, prepared by the Air Transport Association (http://www.air-transport.org), called *ATA SmartBrief,* which offers news summaries and provides links to a variety of news publications that comment on trends in the industry.

2. An indispensable survey of aircraft builders is Bill Gunston, *World Encyclopedia of Aircraft Manufacturers* (Annapolis, Md.: Naval Institute Press, 1993). Donald M. Pattillo focuses on the airframe builders and includes commentary on the difficulties of McDonnell Douglas in *Pushing the Envelope: The American Aircraft Industry* (Ann Arbor: University of Michigan Press, 1998), 338–39, 346, 359–64. For background and incisive insights on the Airbus/McDonnell Douglas/Boeing rivalry, see *Birds of Prey: Boeing vs. Airbus—*

A Battle for the Skies (New York: Four Walls Eight Windows, 1997), by Matthew Lynn, a British financial analyst. Periodical sources for the merger with Boeing, European ambitions, and the growing challenge to American leadership in airliner production include James R. Carroll, "Aerospace Companies Announce Merger," *Austin-American Statesman* (12 December 1996); Richard Aboulafia, "Airbus Pulls Closer to Boeing," *Aerospace America* 38 (April 2000): 16–18; "A3XX: Giant of the Skies," *Aerospace International* 28 (September 2000): 16–19; and Paul Mann, "Labor, Industry Clash over Aircraft Offsets," *AWST* 153 (11 December 2000): 48–49. On the European consolidation in particular, see Anthony Velocci Jr., "Europe Surmounting Hurdles to Become Tougher Competition," *AWST* 149 (3 June 1996): 51–54; John Morocco and Pierre Sparaco, "Franco-German Merger Will Spur Airbus Overhaul," *AWST* 152 (18 October 1999): 26–27; and "EADS Launch Pad," bound in as an advertiser-sponsored supplement, *AWST* 153 (24 April 2000), S1–S24. The following articles were accessed through *ATA SmartBrief:* Peter Robison, "Boeing Engineers Take on Airbus Complaints," *Bloomberg News* (12 December 2000); Kevin Done, "Building the Superjumbo," *Financial Times* (1 November 2000); and "Boeing Bets on Future with Smaller Jets," *Seattle Times* (11 December 2000).

    3. The American Institute of Aeronautics and Astronautics publishes *Aerospace America,* a monthly journal with a strong technical bent. However, it frequently includes highly knowledgeable assessments of key trends, such as James W. Canan, "Fighters Vie for Future Markets," *Aerospace America* 36 (January 1998): 26–33. See also Richard Aboulafia, "Europe's Fighter Renaissance," *AWST* 153 (17 January 2000): 15–18; Gareth Corsi, "X- Rated," *Aerospace International* 28 (6 June 2000): 26–29, which analyzes the American/European competition and NASA's role in advanced programs; and William B. Scott, "Joint Strike Fighter Balances Combat Prowess, Affordability," *AWST* 149 (3 August 1998): 51–57. Paul Mann, "Labor, Industry Clash over Aircraft Offsets," *AWST* 153 (11 December 2000): 48–49, describes a complex trade issue. Alexander Nicol, "BAE Seals Lockheed Deal," *Financial Times,* in *ATA SmartBrief* (29 November 2000) notes international defense issues. For discussions of the UAV phenomenon, see Steven J. Zaloga, "Conflicts Underscore UAV Value, Vulnerability," *AWST* 153 (17 January 2000): 103–4, 106–13; and David Fulghum, "Global Hawk Snares Big Break," *AWST* 153 (23 October 2000): 55.

    4. The growing tendency for many corporations and their accountants to view executive aircraft as useful business tools and not merely perquisites of capitalist barons is examined in John Huey, "The Absolute Best Way to Fly," *Fortune* 129 (30 May 1994): 121–28. Multiple ownership is explained in Anthony Velocci, "Outlook Robust for Fractional Jet Sales," *AWST* 152 (18 October 1999): 40–41. The robust market for general aviation aircraft is described by Ian Sheppard and Marin Wagland, "NBAA," *Aerospace International*

27 (November 1999): 14–16; and Richard Aboulafia, "Business Jet Makers Reaping Strong Market Benefits," *AWST* 153 (17 January 2000): 93–94. On helicopters, see Richard Aboulafia, "Helo Industry: Changes Pending," *AWST* 153 (17 January 2000): 77–78, 80. A survey that touches on virtually all of the most active manufacturers is Donald Pattillo, *A History in the Making: 80 Turbulent Years in the American General Aviation Industry* (New York: McGraw-Hill, 1998).

5. William E. Burrows, *This New Ocean: The Story of the First Space Age* (New York: Random House, 1998), is a rousing story by a veteran journalist. Burrows blends the story of NASA's space program with key projects undertaken by national defense agencies, ending with the politics of the International Space Station and the pioneering probes to Mars. Bryan Burrough, *Dragonfly: NASA and the Crisis Aboard Mir* (New York: HarperCollins, 1998), represents a cautionary view of the American-Russian experience. NASA's new contractual arrangements were described by Mark Carreau, "NASA to Cut Costs with Consolidation Contract," *Houston Chronicle* (26 September 1998). Cogent commentaries on aspects of launch vehicles and various missile programs are presented by Marco Antonio Caceres, "Industry Faces Launcher Excess," *AWST* 153 (17 January 2000): 135–36; and Steven Zaloga, "Missile Markets Stabilizing," *AWST* 153 (17 January 2000): 167–69. See also "Business is Good," *AWST* 153 (11 December 2000): 27, on Boeing's activities and satellite sales. Further commentary includes Marco Antonio Caceres, "Satisfied Demand, Financing Woes Soften Satellite Market," *AWST* 153 (17 January 2000): 151–52. For informative overviews of two major players in the telecommunications business, see "All Eyes on Globalstar," bound as a "Space Business" supplement in *AWST* 152 (4 October 1999): S1–S26; and "After Iridium: Fallout for the Space Industry," bound as a "Space Business" supplement in *AWST* 153 (3 July 2000): S1–S28. For samples of mounting criticism of NASA, see Stanley W. Kandibo, "ASME, NASA at Odds over Aeronautics," *AWST* 153 (24 April 2000): 27–28; and "Editorial: What's at Stake in U.S. Aeronautics Decline," *AWST* 153 (2 October 2000): 82.

CHAPTER 1
The Flamboyant Years 1900–1919

AROUND THE TURN of the century, Orville and Wilbur Wright of Dayton, Ohio, began looking for a spot to test some gliders that they had designed themselves. They settled on Kill Devil Hills, North Carolina, a remote windblown range of sand dunes along the Atlantic coast. Although a long way from home, it suited their needs beautifully. Breezes sweeping in from the ocean provided a steady source of lift for their gliders during some three years of flying experiments. The isolated location also yielded the privacy so dear to the reticent brothers.

In the months between their annual sojourns to Kitty Hawk, the Wrights attended to their thriving bicycle business and began to work on designs for a powered airplane. With the assistance of their skilled machinist, Charles E. Taylor, they designed and built their own engine from scratch, and also fashioned their own propellers. In the fall of 1903, they made their fourth annual sojourn to Kitty Hawk, shipping their aircraft by train. On December 14, 1903, they flipped a coin to determine which brother would make the first flight. Wilbur won the toss but had only just gotten into the air and covered about 112 feet when the plane apparently stalled, hit the ground, and damaged the rudder. Still, the plane seemed to have plenty of power, and the controls seemed adequate. The boys telegraphed their father the next day: "Success assured, keep quiet."[1] During the next three days, the Wrights finished repairs and waited for good weather. Orville's turn at the controls came on

December 17, 1903; the plane lifted into the air and covered some 120 feet during its 12-second journey before Orville landed. The brothers immediately made three more flights, including one by Wilbur that lasted close to a minute and covered 852 feet. The Wrights had achieved the first sustained, controlled flight by humans in a powered airplane. The brothers sent another telegram to their father in Dayton, confirming their success and promising to be home for Christmas. Following their own earlier, terse admonition to "keep quiet," they did little to publicize their achievement for the next four years.

The Wright brothers, aware of patent disputes involving other inventions, wanted time to assure their own patent rights and to perfect various aspects of their airplane. Despite occasional reports of their success, the American public remained skeptical about people flitting about the skies in flying machines. Also, the Wrights needed to determine how the airplane might be useful.[2] In short, they needed to define the market.

The Wrights and the Aviation Business

The market issue became a consistent theme in aviation and aerospace development. In the years prior to World War I, airplanes were hard to sell because of their relative expense, limited payload, and inability to operate in poor weather—a factor that represented a major hindrance through World War II. Moreover, public skepticism remained a severely limiting factor. There was a paradox here. The drama of flight undeniably attracted tens of thousands of spectators to various air shows over the years, and thousands more read a growing number of news stories and books on the subject. But aviation was seen as an incredibly dangerous activity; it was the lurking element of tragedy that drew many of the spectators. Henry. H. "Hap" Arnold, who commanded the United States Air Force in World War II, visited the Belmont, New York, air meet in 1910. A young lieutenant at the time, he recalled the ghoulish expectations of many onlookers: "The crowd . . . gaped at the wonders, secure in the knowledge that nowhere on earth, between now and supper-time, was there such a good chance of seeing somebody break his neck." The problem of safety dogged the young airline passenger business through the 1930s, a fact that prompted the equally young aviation industry to find markets in the American military sector and the military export business.

The Wrights began their experiments in flight as a search for the solution to a challenging technical problem. Prior to their success, they seem to have given little thought as to how flying machines might actually function in modern society. Within a year of their flight at Kitty Hawk, however, the practical uses of aviation began to draw their attention. Octave Chanute, a well-known and financially successful engineer, had been a correspondent of the brothers for some years. Born and trained in France, Chanute became an important link between the aeronautical communities in Europe and America. Late in 1904, Wilbur wrote to him about their continuing development of the Flyer, and confided that "it is now a question of whether we are not ready to begin considering what we will do with our baby now that we have it."

As realists, the brothers knew the limitations of their frail aeroplane, especially given the state of the art in 1904. Early planes flew at 30–40 MPH, and even moderate gusts were enough to convince the Wrights and their pre–World War I contemporaries to remain safely on the ground. Payload was severely limited; the Wrights began carrying a second passenger in 1908, but there was certainly no capability to take off with any additional weight. Still, there was opportunity for improved performance that might make airplanes practical. The Wrights decided that the most likely uses included military reconnaissance, exploration, transportation of passengers and light cargo, and sport. Considering the payload of their present aircraft, aerial reconnaissance seemed by far the most feasible application. "It is therefore our intention to furnish machines for military use first, before entering the commercial field," Orville decided, "but we reserve the right to exploit our invention in any manner we think proper."[3] As it happened, the first significant contract did, indeed, come from military sources.

The Wrights themselves had attempted to contact military officials in 1905, only to receive indifferent responses. Over the next two years, influential contacts pushed the matter in Congress, until the U.S. Army Signal Corps officially asked for a proposal from the Wright Cycle Company and solicited other public bids for a working flying machine. But the Wright proposal was the only one with clear promise. In February 1908, the War Department and the Wrights signed a contract for $25,000 calling for an airplane capable of speeds of 40 miles per hour, a range of 125 miles, and the ability to carry one passenger. All these requirements were to be demonstrated during flight tests at Fort Myer, Virginia, in the late summer.

During September 1908, Orville dazzled observers at Fort Myer as he put the Army plane through its paces and set several new flying records. Then tragedy struck during a flight with Lieutenant Thomas Selfridge. A faulty propeller severed some bracing wires, and the resulting crash killed Selfridge and put Orville in the hospital for several weeks. The accident served as a grim reminder that aeronautical progress would indeed be costly.

In 1908, Wilbur Wright demonstrated the biplane that he and his brother had built to meet requirements set by the United States Army. The flights at Fort Myer, Virginia, generated considerable public interest and won the government's first contract for a military airplane. *Courtesy of the U.S. Navy.*

After an analysis of the crash and its cause, Orville was back at Fort Myer in June 1909 to finish the acceptance trials. On August 2, 1909, the U.S. Army Signal Corps bought its first plane; the Wrights not only got their $25,000 but also picked up an extra $5,000 for exceeding specified requirements for the plane's performance. The faint outlines of an aviation industry began to appear.

The early business negotiations of the Wrights are difficult to sort out, and a number of agreements seem to have been made under the banner of the Wright Cycle Company. The earliest business feelers actually came from the British, who sent an emissary to Dayton in 1904. Drawing information from articles and correspondence of Octave Chanute, the British took the Wrights very seriously. But the Wrights wanted the United States to have the first opportunity to buy their plane, which led to their early correspondence with the American government and the Fort Myer trials. From 1904 through 1907, they had kept their European contacts alive and used one of their French correspondents to sound out the French Army. The brothers were already busy filing patents in America as well as numerous European countries. An investment bank, Charles R. Flint and Company, worked out an agreement to represent the brothers abroad, where Flint had already sold American technology and military supplies, including naval ships to Brazil, cars in France, and a submarine to Russia. Through Flint, the Wrights were soon involved in detailed contract negotiations in Britain, France, Germany, Belgium, and elsewhere.

In the meantime, Wilbur sailed for Europe in the spring of 1907, joined by Orville in July. Along with their financial representatives, they entered the political and financial maelstrom of Continental finances. The U.S. Army, nervous about so much foreign interest, began to have second thoughts and decided to conduct aeronautical trials. The Wrights returned to America late in 1907 and negotiated their Army contract. The following summer, when Orville headed for Fort Myer, Virginia, to fly for the American government, Wilbur headed back to Europe.[4]

Across the Atlantic, many aviation figures regarded the Wright brothers' reputation with polite skepticism. In 1906, Alberto Santos Dumont, a Brazilian émigré living in Paris, had become the first person in Europe to construct and fly a powered plane. The number of European builders and pilots grew impressively, leading many on the continent to discount the achievements of the Wright brothers. During the first week of Wilbur's flights, in August 1908, the French were dazzled. A handful of onlookers grew into massive, awestruck crowds as they watched Wilbur climbing and wheeling through crisply executed figure eight maneuvers. Newspapers throughout Europe wrote hugely flattering news stories. Many leading personalities flew with him as passengers—reporters, royalty, and several women—as Wilbur became the sensation of Europe. Early in 1909, Orville and their sister Katherine joined Wilbur in France. The trio was immediately lionized by the Continental press. "The Wrights," wrote biographer Tom Crouch, "were the first great celebrities of the new century."

All this led to a lively tempo of direct orders for Wright planes, licensed production contracts, stock agreements, and royalties on Wright designs sold overseas, from Norway to Turkey. With aeronautics celebrated as an international phenomenon, attended to by royalty, social lions, and international reporters, and with a series of foreign contracts, serious investors in the U.S. began to take note. The first aviation manufacturing companies in America were about to take shape.

Cash from various prizes in America and abroad, the U.S. Army contract, plus fees from foreign licensing came to a total of a quarter of a million dollars for the Wrights. Late in 1909, the brothers agreed to accept the proposal of a group of well-connected American investors to create the Wright Company, receiving an additional $100,000 in cash with the promise of a 10 percent royalty on every plane and one-third of the shares. From headquarters in New York City, the company would handle all patent suits and legal fees, leaving the brothers personally free of such concerns. With August Belmont, Robert Collier, and Cornelius Vanderbilt on the board, along with Fred Alger of the Packard Motor Car Company, the brothers had clearly captured the business community's attention. The Wright Company was incorporated in New York on November 22, 1909, with a capital stock of $1 million. Wilbur and Orville were named as president and vice-president. The future looked very promising.[5]

Although the European arrangements may have seemed gilt-edged at the start, business from that quarter became mired in design changes and patent disputes, so that sales from the European market remained a disappointment for several years. During 1910, the Wright Company hired a workforce and moved into a new factory in Dayton, with plans to build two flying machines per month. As late as 1915, the company had sold only 14 planes to the U.S. Army, and civil sales limped along. The company buttressed these sales from fees charged to train more then 100 aspiring aviators, including Hap Arnold, the future Air Force general; other significant revenue resulted from aerial exhibition appearances. As president of the Wright Company, Orville took little interest in expanding its manufacturing or long-term financial development. Grover Loening, a graduate of Columbia University with a degree in engineering, became the company's chief engineer in 1913. Loening remained perplexed—and frustrated—by the lack of organization of the Dayton factory. Orville, he said later, was simply not a manager. Moreover, Orville remained disconsolate over the death of Wilbur in 1912, victim of complications from a recurrence of typhoid fever.[6]

Still, Orville was a shrewd enough businessman to plan a profitable exit from the company that carried the brothers' name. During 1914, he bought up all the shares held by the original board of directors. Then he offered the entire company for sale. Although there had been manufacturing problems, assets included the factory in Dayton and the aura of the Wright name. In the autumn of 1915, a syndicate of New York financiers paid an estimated $1.5 million for the company, plus a retainer of $25,000 to claim Orville's services for a year. Orville made good on his contracts but spent most of his time completing his mansion and estate in Dayton, and tinkering with his own personal projects. The reorganized Wright Company was succeeded by a series of mergers and divestitures through the 1920s. One of the deals included formation of the Wright Aeronautical Company (1919), which became a major engine manufacturer in the postwar decades. Aside from occasional consulting, Orville had little to do with any of these developments. He served on the National Advisory Committee for Aeronautics, the Guggenheim Foundation for the Promotion of Aeronautics, and other organizations, but never assumed a leadership role.[7] Nonetheless, at his death in 1948 (age 76), he had become an internationally revered figure in aeronautics.

Early Entrepreneurs

Obviously, other aviation pioneers were active during the prewar and wartime era. Their experiences often had similarities to those of the Wrights, including the significance of foreign markets and the role of military sales as factors in the development of their businesses. The increase of competition, especially in the case of the Wrights and Glenn Curtiss, not only stirred major controversy but also contributed to events that set the stage for a trade association, marking the beginnings of institutionalization in the nascent industry.

Born in Hammondsport, New York, in 1878, Glenn Curtiss grew up in Rochester, spending his summers with grandparents back in Hammondsport. As a resident of Rochester, Curtiss worked for Eastman Kodak assembling cameras. During his Hammondsport summers, he became a champion bicycle racer and eventually set up his own bicycle shop, with agencies in two nearby towns. Curtiss continued to race, and his search for speed led him to experiment with small gasoline engines that he adapted to his cycles. Inevitably, Curtiss began manufacturing his own engines and complete motorcycles; eventually the Curtiss Manufacturing Company sold its machines coast to coast. In California, balloonist Tom Baldwin saw a Curtiss motorcycle in 1904. Determined to adapt such an engine to propeller-driven airships, Baldwin headed east to Hammondsport, where the adventurous Curtiss became an enthusiastic ally. In 1905, the *California Arrow* became the first Baldwin airship to mount a lighter, specially designed engine, representing Curtiss's first venture into flight technology. The success of Curtiss-engined airships caught the U.S. Army's attention, leading to a demonstration for the War Department. During the summer of 1908, the year before the Wrights delivered their first plane to the Army, Curtiss and Baldwin carried out flight trials at Fort Myer, Virginia, meeting Army specifications for a flight duration of two hours and top speed of 20 MPH. Consequently, the Curtiss-Baldwin airship became U.S. Army Signal Corps Dirigible

No. 1 on August 18, 1908, marking the origins of the military as a customer for powered-flight technology.[8]

The resultant publicity brought Curtiss to the attention of the inventor Alexander Graham Bell. At Bell's laboratory at Baddeck, Nova Scotia, Curtiss collaborated in the design of several airplanes using his engines. Under the umbrella of this Aerial Experiment Association, two airframes were built at Hammondsport and christened *Red Wing* and *White Wing* for the color of their fabric covering. There were additional aircraft and several record flights before Curtiss decided to part with the Aerial Experiment Association in 1909 and reorganize his own business interests to include the Curtiss Aeroplane Company, announced in December 1911. (Curtiss had already begun to build aviation motors and advertise them. As early as 1908, one advertisement proclaimed: "Curtiss Motors/Especially Designed for Aeronautical Purposes. . . .") Curtiss's association with the renowned Alexander Graham Bell gave him the sort of stature that led interested buyers to seek him out. In this context, the Aeronautic Society of New York approached Curtiss with a request to build an airplane for the use of this group of wealthy gentlemen, to include flight instruction. The agreed-upon price came to $5,000 for the airplane, called the Curtiss No. 1, delivered on May 29, 1909, only one week late. In many ways, this contract and delivery of an airplane to a discrete buyer represents the origins of the civil aviation manufacturing industry. In the wake of sensational publicity following the Rheims aerial exhibition in France during 1909, the Curtiss operation in Hammondsport picked up orders for an estimated eight new planes. Moreover, at least one of the sales came through a distributing agent, Wyckoff, Church, and Partridge, an automotive dealer in New York City who set up a special "Aeronautical Department" to sell planes and took out full page ads in national journals like *Life* magazine to reach prospective customers. The business of developing, marketing, and selling aircraft had begun to evolve.

By 1913–14, Curtiss's own company had developed a standard line of products, comprising six different models of varying engines (40 to 75 horsepower) and capability, and available with pontoons as a "Hydro-aeroplane Attachment . . . For rising from and alighting on the water." The pontoons were available for $500; the planes ranged in cost from $4,500 to $5,500. In addition, Curtiss developed a successful flying boat hull, and several different models were sold to the U.S. Army, Navy, and various sportsman pilots. Because aircraft like this cost $6,000 and up, buyers typically came from the ranks of wealthy sportsmen like Harold F. McCormick, head of the International Harvester Company of Chicago, who also had a personal pilot to shuttle him back and forth between his

Lake Forest estate and the Chicago Yacht Club. Prospects for an industry built on the civil market were obviously limited. Furthermore, sales to the military involved only a handful of planes.[9] In order to develop a source of steady income and to broaden the market, Curtiss, the Wrights, and others came to rely on aerial exhibitions.

In 1910, the Wrights hired Roy Knabenshue, an experienced promoter of balloon flights, to run the Wright Exhibition Company, hoping that aerial shows might entice adventurous customers to buy airplanes. In any case, the air show business generated a much-needed cash flow. The Wright Exhibition Company charged $5,000 for each plane appearing at an air show. Depending on prizes won and the number of times they flew, leading pilots might earn as much as $6,000 to $7,000 per season. It is not clear how much profit developed from the air show business, but from 1910 to 1912, the Wright Exhibition Company grossed $1 million per year. Clearly, there was money to be had from aviation.[10]

Regrettably, many pilots knew little about aerodynamics and the performance they could realistically expect from the frail aircraft of the prewar era. Frank Coffyn was one of four men who signed a two-year contract to fly with the Wright Exhibition Company, but he was the only one who lived to fulfill it. Sensational stories of aircraft smash-ups made lurid headlines all over the country and prompted a barrage of criticism of aerial antics in newspaper editorials and cartoons. In a series of articles during 1911, *Scientific American* deplored the fatality rate and argued for practical demonstrations of planes as transports and not as sources of "sensational amusement." After two seasons, the Wrights withdrew from aerial exhibitions with distaste, although other aerial troupes continued to operate.

Although the image of danger continued to follow commercial aviation development for several decades, change was already in the wind. In November 1910, a department store in Columbus, Ohio, contracted with the Wrights to fly a bolt of silk from Dayton, a heavily publicized event in which the plane beat a crack express train between the two cities. There were demonstrations of airmail flights around the country in 1911, and promoters in California and Florida experimented with aerial passenger service in the years just before World War I. All of these things hinted at possible markets in the future. Still, many informed observers remained somewhat skeptical that a major industry might emerge out of this. In 1915, Professor Jerome Hunsaker, who taught one of the first courses in aero engineering at the Massachusetts Institute of Technology, received a letter from Glenn Martin in California.

Born and raised in the midwest, Martin had headed for California in 1905 to set up an auto repair shop. Fascination with flying led to a small company to design and build planes, including a contract to produce some trainers for the U.S. Army. With a sudden expansion of production orders and governmental contract specifications, Martin needed an engineer for his fledgling airplane company. Donald Douglas, one of MIT's recent graduates, needed a job and expressed an interest in Martin's company on the west coast. Hunsaker advised Douglas to stick to mechanical engineering. Douglas later recalled Hunsaker's warning as something to the effect that "[t]his airplane business will never amount to very much."[11]

If nothing else, the query from Martin symbolized a persistent, if slow, growth in aviation manufacturing. The American end of the business began to attract dozens more entrepreneurs, including several from abroad, whose legacy to the American experience was a lasting one. Realizing the need to keep up with developments overseas, Glenn Curtiss went to Europe in 1913 and stopped off at the factory of Sopwith Aviation in England. During his visit, a chance conversation with B. Douglas Thomas, a Sopwith engineer, led to an invitation from Curtiss to submit a design for a new tractor biplane. Curtiss liked the proposal and offered Thomas a job. Thomas arrived in America in the spring of 1914. For Curtiss, he designed two biplane trainers and became a leading designer for the notable flying boat *America*, built to fly across the Atlantic. Although the *America's* flight was canceled at the outbreak of World War I, the plane influenced later Curtiss programs, and the trainers evolved into the famous JN "Jennie" trainers of World War I. Douglas Thomas himself joined another early firm and assisted in a respected wartime design, the Thomas-Morse Scout biplane fighter.[12]

Despite the identical name, Douglas Thomas was not related to any of the principals at the firm of Thomas-Morse, although the English connection between them was real enough. The early European influences on American aviation continued in the persons of the Thomas brothers, William and Oliver, who received college degrees from Central Technical College in London in the early 1900s. They came to America, worked for various engineering firms (including Curtiss), then decided to build airplanes as the firm of Thomas Brothers, based in Bath, New York. Using income from exhibitions and from their flying school, they built several different designs. The war in Europe brought enough component orders from overseas to warrant a move to a much larger factory in Ithaca. Also, B.D. Thomas now left Curtiss and joined the brothers as chief engineer and designer. Within a few months, Thomas Brothers had landed an order from Britain for two dozen biplanes, and organized a subsidiary to manufacture engines.

The American government began placing orders for Thomas planes as early as 1916. The need for plant expansion and additional capital led to a merger with a local manufacturer, the Morse Chain Company. As Thomas-Morse Aircraft Corporation, the organization trained many American and Canadian fliers and developed new planes such as the successful Model S-4 and the MB-3. The workforce rose to 100 and then to more than 1,200 employees. During World War I, Thomas-Morse emerged as one of the leading manufacturers in the country. In the early 1920s, Oliver concluded his aviation career and settled in Argentina as a rancher. William also resigned from Thomas-Morse, although continuing as an aeronautical consultant. The Thomas-Morse organization eventually sold out to Consolidated Aircraft Company in Buffalo, New York. During the 1930s, William became a leading figure in model plane building and played a key role in developing the hobby as a national phenomenon. Eventually he became president of the Academy of Model Aeronautics, a group that often served as a point of departure for so many future leaders in the aerospace field.[13]

For every aviation hopeful like William or Oliver Thomas, who enjoyed some success, many more slipped into obscurity. The names of their firms reflected their high hopes, as well as the diversity of businesses who suddenly sensed an opportunity to cash in on the growing aviation craze. From about 1907 on, dozens of airplane companies sprang up, and curious aspirants like the Swivel Buggy and Wagon Company announced their entry into the new field. In 1910, the Scientific Aeroplane and Airship Company notified the public that it was building planes and that one could be had for the price of $5,000. The company required a down payment of one-third (cash), asked three weeks for delivery, and guaranteed the plane to fly. Unfortunately, the big aviation boom failed to materialize, and by the era of World War I, Scientific had disappeared along with such optimistic entrants as the Aerial Yacht Company and dozens more.

Although the majority of companies fizzled, their activities stimulated a market for engines and accessories. In addition to Curtiss and Wright engines, a variety of other power plants were adapted from existing automotive production models, including several from marine engines, such as the Elbridge "Featherweight," a 45-horsepower engine used on Martin, Farman, Curtiss, and other planes, and the Roberts "6-X Engine" of 75 horsepower, used by Benoist, Bleriot, Curtiss, and others. In addition, suppliers like R.O. Rubel, Jr., and Co. of Louisville, Kentucky, began to emerge. Their prewar catalog of 1913 offered a melange of equipment, such as tires, "rubber aeroplane springs," fuel tanks, wheel spokes, round and square steel tubing, fabric, bamboo, spark plugs, turnbuckles, landing gear, and finished wing ribs in either Curtiss style or Bleriot style, the latter available as a full set of 24 at $24.[14]

One of the leading figures of the early years of American aviation, Grover Loening was one of the first engineers with an American college degree to play a role in the nascent aviation industry, going on to become a pioneering manufacturer. For several decades he also served as a consulting engineer and became the first president of the industry's trade association, the Aeronautical Chamber of Commerce of America. Loening was born in 1888, in Bremen, Germany, where his father served as U.S. consul. He received his education in the United States and eventually enrolled at Columbia University, where he received a B.S. degree in 1908 and his M.A. in engineering two years later. The young engineer with a fascination for airplanes found a job with the Queen Aeroplane Company.

The Queen organization was a small one, and survived by building copies of the Bleriot monoplane, made popular by Bleriot's historic flight across the English Channel in 1909. The Bleriot's success in Europe made the Bleriot IX model a sought-after one in America. Although the Bleriot company offered licenses, the majority of Bleriot machines built in the U.S. appear to have been pirated from available plans or copied from various planes that found their way to the U.S.

The Bleriot monoplane, like the Wright biplane, became a stock symbol of the new age of aviation. One Bleriot, suspended on piano wires, thrilled New York audiences at the Astor Theater, where it functioned as the centerpiece of a play billed as *The Aviator*. This sort of popularity made it a candidate for series production in the U.S., a project promoted by Willis McCormick, a sportsman and wealthy New York stockbroker. His firm, the Queen Aeroplane Company, went into business in 1910 and numbered some 85 employees within a year, working in a hodgepodge of buildings that had once housed an amusement park at 197th Street and Amsterdam Avenue. Finished planes ranged in price from $2,900 to $5,000, depending on the engine. No accurate record of production numbers survived, but several dozen aircraft appear to have been fabricated, flown by several leading American aviators of the era. Grover Loening joined Queen in 1911 and learned the fundamentals of aircraft manufacturing.

Loening's degrees from Columbia and practical experience with the Queen company won him a job with the Wright Company in 1913. Although he became general manager of the factory being built at Dayton, he was never happy with frequent corporate squabbles and became increasingly frustrated by Orville's haphazard leadership. The next year, Loening left Wright to take the position of chief aeronautical engineer for U.S. Army Signal Corps and headed for the salubrious environs of San Diego, California. Loening found plenty of work to do in San Diego, where inexperienced student pilots all too often became victims of the fickle Curtiss and Wright training planes. Some planes built by Glenn Martin had nose-mounted engines (called the "tractor" position) that had a much safer flying record. Loening and Lieutenant Thomas Milling made a hurried trip to Glenn Martin's shop in Los Angeles and persuaded him to deliver a tractor trainer, the Martin Model T, on the promise of a contract. Loening's desperate visit became the basis for Martin's success in government contracting. Loening himself found additional sources of income as an aero entrepreneur. Convinced that many accidents stemmed from badly designed engineered gear, he designed a new configuration that eliminated traditional skids and instead put the wheels farther forward with special bracing. Loening eventually received a patent. With great satisfaction, he recalled that Curtiss used it on the famous JN series and paid him royalties for many years.

There were other ways to make money. At San Diego, he began offering informal lectures in airplane design and construction for U.S. Army flying officers at the field. Eventually, the lectures became a pamphlet, and the pamphlet evolved into a book, *Military Aeroplanes*, which Loening published himself after an executive officer designated the books as required reading. After the war started in Europe in 1914, the British government ordered 5,000 copies (which eventually rose to 18,000); American entry in 1917 created a large market at home. At war's end, Loening's book had sold 43,000 copies for nearly $100,000 in profit. At the same time, *Military Aeroplanes* became one of the early textbooks that provided support for the evolving industry.[15]

Patterns of Institutional Development

The success of Loening's book reflected the hunger for information about aviation technology, and the paucity of organized, institutional sources available to provide it. Even before the declaration of war in Europe, military aviation developments overseas clearly overshadowed those in America. A number of concerned individuals decided to rectify the situation, and, among other achievements, came up with a reasoned analysis of the prewar European programs that appeared to have so successfully advanced aviation development across the ocean. Activities to boost America's aviation capability thus occurred against a background of public opinion that sounded the alarm about the lack of American military acumen, including the nascent field of aeronautics. Consequently, the evolution of aviation institutions took place as a mix of actions taken by alarmist groups, federal legislation, business issues, and the dictates of war itself.

Under the auspices of the Smithsonian Institution, two American experts arrived in Europe during 1914 to assess the situation overseas: Dr. Albert Zahm taught physics at Catholic University in Washington, D.C., where he also conducted aeronautical experiments; and Dr. Jerome C. Hunsaker, a distinguished graduate of the Massachusetts Institute of Technology, who was planning to return there to develop some early courses in aeronautical engineering. Their travels in Europe established contacts that later proved valuable in promoting theoretical research through federal agencies in America. Meanwhile, their report, issued in 1914, emphasized the galling disparity between European progress and inertia in the United States. The outbreak of World War I in Europe served as the catalyst for the creation of an American agency as a matter of national security.

In 1915, while drafting the charter for what became the National Advisory Committee for Aeronautics, Hunsaker and the others on the drafting committee drew heavily from European sources, especially Great Britain. The NACA's famous directive, to undertake "the scientific study of the problems of flight, with a view to their practical solution," was taken, word for word, from the first report of the British Advisory Committee for Aeronautics for 1909–10. The British committee also served as a model for the NACA in formulating the membership of its principal committee and in establishing initial areas of aeronautical research.[16]

The legislation for the NACA slipped through almost unnoticed as a rider attached to the Naval Appropriation Bill, on March 3, 1915. It was a traditional example of political compromise. The legislation did not call for a national laboratory, since President Wilson apparently felt that such a move, taken during wartime conditions in Europe, might compromise America's formal commitment to strict nonintervention and neutrality. The proposal emphasized a collective responsibility through a committee that would coordinate work already under way. The committee was an unpaid panel of 12 people, including two members from the War Department, two from the Navy Department, one each from the Smithsonian, the Weather Bureau, and the Bureau of Standards, and five more members acquainted with aeronautics.

For fiscal 1915, the fledgling organization received a budget of $5,000, an annual appropriation that remained constant for the next five years. This was not much even by standards of that time, but it must be remembered that this was an advisory committee only. Once the NACA isolated a problem, its study and solution was generally accomplished by a government agency or university laboratory, often on an ad hoc basis with limited funding. In a wartime environment, the NACA was soon busy. It evaluated aeronautical queries from the Army and conducted experiments at the Navy Yard; the Bureau of Standards ran engine tests; Stanford University ran propeller tests. As the war ended, civil and federal authorities won approval for the NACA's own laboratory, located at a new U.S. Army airfield across the river from Norfolk, Virginia. The military facility was named after Samuel Pierpont Langley, former secretary of the Smithsonian; the NACA facility was named Langley Memorial Aeronautical Laboratory, soon shortened to the familiar, cryptic "Langley" after its dedication on June 11, 1920.[17]

In addition to federal entities like the NACA, other institutional trends began to emerge. As early as the outbreak of World War I, special interest groups appeared, generating publications and publicity intended to promote both national preparedness and the aviation business. Once the battle had been joined, posters reflecting that unquenchable Yankee optimism began to appear, showing skies crowded with American planes, icons of faith in American productivity. Wartime planners assumed that the mass-production techniques developed for the auto industry would quickly be adapted to airplanes. A group of seven auto engine firms agreed to produce 22,500 new Liberty engines; other officials confidently predicted that U.S. plants would deliver so many planes that we could also plan to supply our allies' needs by 1918. *Leslie's Illustrated Weekly* caught the flavor of this enthusiasm in its cover for November 10, 1917, in which an American gladiator, astride a Liberty-engined warplane, is carried into victorious battle against the enemy. Despite America's inability to achieve the great production records that had been forecast, the promise of this capability directly influenced Germany's actions. Following the U.S. entry into the war, German planners launched a strategy of mass production, called the *Amerikaprogramm*, designed to blunt the impact of America's manufacturing prowess.[18]

The mixed record of American production during the First World War was due to a combination of many factors, including the timely resolution of a pernicious patent dispute involving the Curtiss interests and the Wright brothers. The dispute was a long, complex, and bitter conflict involving arcane principles of ailerons and aircraft control. The Wrights had filed their first patent in 1903, although the official U.S. patent was not entered until 1906. Implicit was the idea that all subsequent aviation builders would need to pay royalties to the brothers Wright. Research and experimentation might be allowed, but public exhibition flights and commercial uses fell under their patent rights. During 1909, as Curtiss built new planes, announced sales, and collected headlines in public air shows, the Wrights filed formal suits. The patent skirmish became a vindictive running battle, one that Grover Loening remembered as consuming so much of Orville's energies that he had little time left to keep up with the competition.[19]

The patent wars divided the aeronautical community in America and probably hindered the progress of U.S. aviation. European aircraft began performing to standards that the U.S. could not match. True, the European firms received subsidies, and international rivalries pushed development overseas. But the patent issue tended to discourage bold innovation in America. In the spring of 1917, when America became involved in World War I, the Wright patents, as well as subsequent patents filed by other manufacturers, threatened to undermine the nation's ability to produce large numbers of planes for the war effort. Under pressure from several federal agencies, the bickering aircraft companies formed the Manufacturers Aircraft Association in July 1917. The association's members hammered out a cross-licensing agreement, a common pool of patent fees, and a system of equitable fee payments. The association thus represented an important forward step in the evolution of the aircraft industry.[20]

Some of the wartime institutional trends, including the role of auto manufacturers, became a source of friction with early aviation firms. Loening viewed the formation of the American Society of Automotive Engineers in 1916 as part of the "Detroit conspiracy" to maneuver aircraft manufacturing into the hands of car manufacturers. "Here is where many of us in aviation smelled a rat," Loening claimed, but they were not successful in heading off the pervasive influence in wartime production by the auto industry.[21]

Aviation Production in Wartime

In the meantime, the production of planes in the U.S. began to accelerate. Throughout the prewar years, the vast majority of aircraft "manufacturers" turned out only one or two flying machines each, but the cumulative work of a few stalwart builders began to pay off in terms of orders. Success of American aerial demonstrations overseas, plus the growth of international tensions in the years prior to World War I, led to more orders from abroad. The value of aeronautical exports more than doubled in two years, from $100,000 in 1912 to $226,149 in 1914, with deliveries of planes and engines to Russia, Romania, Japan, and Mexico. In 1914, 49 planes were sold, and 34 went to the export market. As World War I escalated, British, French, and Russian orders accelerated. In 1916 alone, America built 411 planes, exported 269 of them, and collected $2,185,000 from foreign buyers.[22]

After war broke out in Europe in 1914, the belligerents launched an intensive effort to acquire more warplanes. As a means of facilitating large orders from Canada and Britain, the Curtiss Aeroplane and Motor Corporation quickly established a manufacturing unit in Toronto, Canada—Canadian Aeroplanes Limited—which delivered 1,200 training planes by the end of the war. Meanwhile, other Curtiss products went to Britain and Russia. The British ordered some 18 large flying boats, costing up to $24,000 apiece, and built several more under license. Curtiss also landed a substantial contract with Russia for 54 Model K single-engine flying boats, delivered during 1915 for a total of $1 million. The company sent test pilots to complete acceptance flights for final delivery, and those hardy souls traveled by steamship from New York to Archangel, then by train to Sebastopol, on the Black Sea. A number of the planes were shipped across the Pacific to Vladivostok, where they were loaded on the Trans-Siberian Railway and sent westward. The Curtiss test pilots had a difficult time assembling and flying the planes for acceptance, since some parts proved to be faulty. There were other nagging problems with rusted metal fittings and wires, the result of having been shipped underneath numerous crates of fish, the salty brine from which had leaked all over the airplane components.[23]

An accurate assessment of the scope of the American aviation business in these early years remains contradictory and confusing, since much of the information seems to have arisen from anecdotal sources and informal records. Government statistics for 1914 suggest the value of aviation products at $790,000, a pittance in comparison to each of the top 56 industries whose gross value of products was $100,000,000 or more in 1914. As for civilian production alone, the classification "airplane" did not appear in census figures until 1912, since aircraft in the previous years had been classified with "all other cars, carriages, and vehicles." At any rate, federal records specified the production of only 39 planes in 1912; the total of 411 planes produced in 1916 represented a rapidly growing capacity that military planners in Europe eyed with great anticipation.[24]

When the United States entered the war in 1917, the Allies naturally expected a boost from American personnel and production potential, and America confidently expected to meet the demands and still do more than its share. To satisfy foreign requests and the needs of the United States as well, the Joint Army and Navy Technical Aircraft Board called for 8,075 training planes and 12,400 service planes. The production plan for this total of 20,475 planes was to be met in 12 months. The accompanying engine estimate called for 41,810 power plants the first year and 6,159 per month thereafter. This from an "industry" that only the year before had produced all of 411 planes.

Congress appropriated $640 million to do the job, the largest single appropriation for a specified purpose ever legislated in the United States.

An awed public expected to see vast fleets of aircraft crowd the skies. The situation on the ground, however, was anything but heartening. Few of the companies had ever produced anything more advanced than a training plane, and there were only a handful of designers capable of that job. Some car builders had experimented with airplane engines, but the situation in the automobile industry was little better. As for instruments, most of the dials, gauges, and other accessories required for airplanes were completely unfamiliar to American manufacturers. A new industry capable of producing such equipment had to be brought into being, along with methods found to fabricate sensitive instruments on an assembly-line basis. A multitude of lesser problems remained to be solved. There was spruce to be cut for wing spars and yards of fabric to be obtained for covering airframes. There was the problem of lubrication. Ordinary motor oil froze at high altitudes, and the best substitute was castor oil, requiring the U.S. to import a shipload of seeds from India in order to cultivate enough castor plants to satisfy projected requirements.[25]

During the hectic scramble to solve these and other difficulties, the aviation industry fell prey to widespread graft and corruption as companies and managers formulated mergers, interlocking directorates, and anything else to get a share of multimillion-dollar war contracts. On the other hand, as the embryonic aircraft industry struggled to meet the demands, some extraordinary engineering feats were accomplished. Production peaked in 1918, with deliveries of nearly 14,000 airplanes. By 1919, 31 companies were producing aircraft and parts valued at $14 million. All things considered, an industry worth less than a million dollars only five years earlier had made remarkable progress.[26]

But the most notable American production record involved the Liberty engine used by the hybridized DH-4. In May of 1917, J.G. Vincent, of the Packard Motor Car Company, and E.J. Hall, of the Hall-Scott Motor Car Company, took over a hotel room in Washington for nearly a week and designed the 8– and 12–cylinder Liberty power plants with mass-production procedures in mind. On June 4, 1917, the Aircraft Production Board authorized construction; final design, manufacture, and assembly of the first 8–cylinder version was completed in the astonishingly short span of less than six weeks. More than 13,000 engines came off the assembly prior to the Armistice; more than 20,000 were built by the time wartime production ended early in 1919. As military surplus, the Liberty-powered DH-4 planes, thousands of Liberty engines, and JN-4 trainers would all play a utilitarian role in the evolution of American aviation during the postwar era.

Still, most American aerial units overseas flew into combat in French Spads and Nieuports. The American Expeditionary Forces fighting in Europe had some 6,300 aircraft at their disposal, but only 1,200 from the

United States—mostly British DH-4 aircraft built under license and considered by many to be obsolete by the end of the war. In short, the record of American production for combat aviation proved to be mixed. The American aircraft industry employed 175,000 personnel, and the industry claimed to have reached a potential production rate of some 23,000 planes per year. By comparison, the French reported a production capacity of 31,000 planes per year at the Armistice and the British rate had reached 41,000 per year in July 1918.[27]

As these figures made the rounds of national newspapers and congressional committees, the presumed stature of the American industry began to shrink. Coupled with horror stories of rampant profiteering and gross incompetence, the search for culprits intensified. Elsbeth E. Freudenthal's book, *The Aviation Business: From Kitty Hawk to Wall Street*, published in 1940, meticulously recounts dozens of incidents of economic malfeasance that surfaced during wartime and early postwar investigations. Unfortunately, too many optimistic planners had completely misunderstood the complex and fragile nature of aircraft construction. Automobiles, with a rigid steel chassis and comparatively slab-like metal panels forming the body, could be readily assembled and bolted together by unskilled workers. Airplanes of the era were compounded of wood, glue, screws, fabric, and carefully adjusted bracing wires, all requiring the careful ministrations of workers who were more like craftsmen than unskilled laborers. Airframe construction simply was not a procedure that lent itself to production peaks within the few months that American managers allowed for the task.

In the aftermath of seamy trials, federal hearings, and sensational headlines, Congress determined not to allow such shenanigans to recur, especially in the aviation industry. Curiously, it was not clear that Congress—or any government agency—possessed the sort of wisdom to compile a better record than the villains presumed to have held sway in the aviation business. Nonetheless, as a result of the furor, the postwar military aviation industry experienced considerable congressional oversight.[28]

From the earliest years of aviation development, European influences played a significant role, beginning with Britain's query to the Wright brothers in 1904. Similarly, military sales became a goal of early aircraft builders. Each of these factors played a role in the evolution of export sales, even before the United States entered World War I. At the same time, countless individual entrepreneurs and companies tried to enter the alluring business of aviation, despite efforts by the Wright interests to dominate aircraft construction through patents. Glenn Curtiss, Grover Loening, and others represented a rich source of entrepreneurial spirit that helped spread the practice of aeronautics throughout the country and attract a widening spectrum of investors.

Concurrently, the formation of the National Advisory Committee for Aeronautics—again with European influence evident in its charter—marked the growth of an institutional infrastructure. Institutional development offered research support and provided an environment for the growth of interest groups, journals, and related factors to develop and spread aviation knowledge. World War I concentrated all this activity and put aviation in the intense spotlight of national pride. Failing to achieve the grandiose production records that had been expected, the American aviation industry still greatly increased its production capabilities, and saw the emergence of a nucleus of people and business groups who expected to create successful ventures in the postwar era. All of this set the stage for postwar conglomerates like the Curtiss-Wright Corporation. Formed in 1929 after both adversarial namesakes had withdrawn from corporate activities, Curtiss-Wright represented the precedence of hard-headed postwar financiers over prewar pioneers.

CHAPTER 2
Outlines of an Industry 1919–1933

IN 1927, CHARLES A. Lindbergh completed his historic solo flight from New York to Paris in a single-engined monoplane. One of the most memorable events of the 1920s and 1930s, Lindbergh's achievement stimulated widespread interest in aviation and triggered a surge of investments in aviation businesses and stocks. It is important to remember that Lindbergh's flight was not so much a beginning as it was representative of a particular level of achievement. Other planes had crossed the Atlantic in 1919; a trio of single-engine biplanes had circled the world in 1924. What made Lindbergh's flight so newsworthy was the publicity and prize money that attended it. The real aeronautical significance of the event was to be found in the effective, efficient design of Lindbergh's monoplane and the reliability of its radial engine that ran for over 33 hours. The basic technology for success had been in place several years before the flight; the rapid progress of the late 1920s and 1930s was rooted in these prior developments. Moreover, Lindbergh's solo flight, by itself, had little practical effect on the lives of millions of Americans. The practical uses of aircraft in long-range transport for passengers remained in a state of evolution.

An aviation industry needed a consistent market, something that developed slowly in the 1920s. The experience of World War I encouraged aeronautical partisans to hope that aviation had demonstrated its potential for civil development as well as its importance for national security. But in order to translate these assumptions into sales contracts, aviation still had to demonstrate its utility in the postwar world. Particularly in the case of civil aviation, planes had to work for a living and yield a profit in the bargain—a dictum as true for air transport manufacturers as for the airline operators who flew them. The postwar civil industry also included a host of small manufacturers who produced hundreds of compact "light plane" designs for pleasure flying, business trips, and utility operations such as aerial photography. Military aircraft evolved against a backdrop of changing priorities and doctrines. Army aviation began to build a capability for long-range bombing; naval aviation developed in response to such major innovations as the aircraft carrier. For the big manufacturers, military sales embodied the largest source of income by far.

All of these trends occurred in interaction with new institutional entities such as the National Advisory Committee for Aeronautics, a significant source of applied research and invaluable disseminator of information for designers and manufacturers, and the expanding professionalism represented by aeronautical curricula in colleges and universities. Formation of the Aeronautical Chamber of Commerce represented the intention of several early aviation firms to develop a relative agenda for aviation progress. The Air Mail Act of 1925 transferred government air operations to commercial carriers, providing revenues that launched civil air transport companies in the United States. Federal regulation, in the form of the Air Commerce Act of 1926, constituted an important step forward in respect to safety and uniform standards for manufacturers and operators alike. The aviation business, which remained small in comparison to commercial and industrial giants in the automotive field, began to develop the necessary infrastructure to sustain continuing growth and global success. This last factor was critical, because foreign competition often played a significant role in the evolution of America's aeronautical enterprise.

Competitive Symbolism: The Schneider Trophy Races

The American aviation industry, which had evolved largely as the result of the war in Europe, functioned in an international arena. During the 1920s, aviation development often unfolded against an international background of high-speed races and long-range flights, as the aviation community sought ways to demonstrate the potential of this new technology. Often, governments became involved, since aeronautical record-setting was interpreted as a barometer of national prowess in the technological marketplace. Commentary in American trade journals and popular periodicals confirmed the role of aviation in boosting national pride.[1]

Perhaps because seaplanes had a certain cachet that led many people to associate them with the yachting aristocracy, the international Schneider Trophy Races were followed with intense interest. Jacques Schneider, a French industrialist, wanted to promote floatplane and flying-boat development; contests took place at different venues in Europe and America. Begun in 1913 and ending in 1931, the Schneider contest remained a high-profile event and represented an intense quest for speed and national prestige.

American involvement in the Schneider races was brief but significant. After watching from the sidelines, the U.S. finally made a run for the trophy in 1923, encouraged by influential figures who believed that American competition would enhance the design and development of high-performance American fighter planes. Moreover, the international exposure of American technology might bring orders for aviation products as well as planes. Fortunately, several candidate planes were available, including a pair of Curtiss CR-3 floatplanes equipped with Curtiss D-12 engines. The 1923 Schneider race was held in southern England, off the Isle of Wight. The Americans turned in a sensational performance, with the Curtiss biplane racers taking both first and second place. Just a few years earlier, American combat pilots had been forced to fly British and French planes during World War I because the Americans lacked planes with the sort of performance to fight in European skies. Suddenly, the Yankee aircraft had eclipsed the best of the European industry. The British press praised the aerodynamic refinements of the Curtiss racers and their Reed metal propellers. The low frontal area of the Curtiss monobloc engine, said the *Airplane*, was "astonishing."

Just as important, in the opinion of many observers, had been the teamwork and organization evidenced by the U.S. Army and Navy, along with commercial suppliers. The government support of the U.S. effort fundamentally changed the Schneider competition. Corporate sponsorship was no longer enough; the event now became a showcase of advanced technology with planes representing the technical prowess of their respective countries. Despite their successes in 1923 and 1925, the Americans planned to withdraw after the 1926 meet. The prior Schneider victories had not triggered sales of military aircraft, at home or abroad, as U.S. officials had hoped. In any case, Italy took the trophy that year, and the Americans came in second. In retrospect, the U.S. developed usable technology in engine design and aerodynamics.[2] More important, the Schneider competition, involving government support in order to publicize American technology, marked a rite of passage. In the future, the U.S aviation industry realized it would be vigorously challenged by other nations in the international marketplace. Support by the American government often played a significant role in securing contracts overseas as well as promoting a favorable environment at home.

Civilian and Military Customers

With the exception of the Curtiss organization, most of the high-profile companies of the World War I years disappeared in the postwar era or survived under a different corporate banner. Thomas-Morse and others gave way to shifting business alignments and financial factors. Among automotive businesses, only the Ford Motor Company evidenced much interest in postwar aviation. Between 1926 and 1932, Ford produced nearly 200 models of the stalwart Ford Tri-Motor, a 12-passenger plane. Many in the aeronautical field hoped that Ford's impressive industrial reputation and mass-production experience would revolutionize the aviation business. In reality, the Tri-Motor, while nearly indestructible, was not easy to fly and was none too comfortable for passengers. Many of its construction features, arbitrarily incorporated into the design to facilitate Ford's mass-production practices, contributed to nagging aerodynamic shortcomings. Faltering sales during the Depression and pressing issues in other sectors of Ford's industrial empire led to an abrupt decision to dismantle the airplane division in 1932. For all of its high hopes, the division had cost Henry Ford close to $6 million in losses. Still, his entry into the aviation field had conferred prestige on a struggling industry and facilitated crucial investments into many new firms. The U.S. industry's postwar leaders proved to be a group of younger companies who emerged from existing enterprise, or found venture capital, or somehow scrambled successfully and survived with outstanding designs. The origins of Boeing, Douglas, Grumman, and Lockheed reflect the diversity of the period.

William E. Boeing presided over a successful lumber operation in the Pacific Northwest in the years prior to World War I. During 1915, as a hobby, he took flying lessons at a school in Los Angeles run by Glenn Martin. He and a friend designed a simple, two-place trainer, and won an order to produce 50 of them after the United States declared war on Germany and Austria in 1917. During the early 1920s, Boeing won additional contracts to build planes for the U.S. Army and Navy, becoming a leading aviation firm by 1927, with about 800 employees. In 1928, the Boeing Airplane and Transportation Company went public, quickly selling its stock to aggressive investors. From the very beginning, the company was somewhat unique, building its own airliners and operating them over its own routes.

Donald Douglas studied at MIT before World War I, where he helped Jerome Hunsaker build a pioneering wind tunnel in 1914, acquiring rare training in formal courses of aeronautical engineering. After wartime work with Glenn Martin in California and Ohio, Douglas returned to the West Coast. In Santa Monica, California, he found local venture capital to set up his own company, which also benefited by investments from Los Angeles *Times* publisher Harry Chandler and a circle of Chandler's wealthy friends. Beginning as a design office in the back of a barber shop, the Douglas Aircraft Company won a series of Army and Navy contracts during the early 1920s and moved into successively larger quarters for manufacturing.

On the East Coast, Leroy Grumman, a U.S. Navy ensign, worked with Grover Loening on early Navy projects through 1920, when he joined Loening Aeronautical as a full-time test pilot. Loening gave him increased design responsibilities over the years; when Loening sold out in 1930, Grumman organized his own firm. Grover Loening and his brother invested $30,000—nearly half of the capital used by Leroy Grumman—to launch the new company. The Grumman Aircraft Engineering Corporation set up shop in an abandoned garage on Long Island. During the early Depression years, Grumman eked out a survival through maintenance and support contracts, until development of a reliable retractable landing gear helped win profitable contracts for fast, rugged U.S. Navy fighters in the mid-1930s.

The evolution of the Lockheed Corporation was somewhat complex. Allan and Malcolm Loughead changed the spelling of their name before World War I. In 1917, in California, the brothers formed the Lockheed Company, which offered flight training and light manufacturing services during the war. They used $1,800 of their own savings, plus $1,200 from an associated taxicab company. The company found little business in a postwar market that was flooded by war surplus aircraft, parts, and engines. Discouraged, Malcolm took a job in the automotive industry, but Allan reorganized operations in 1926 as the Lockheed Aircraft Company, relying on $22,000 from a Los Angeles investor. He rehired a young, self-taught designer, John Northrop, and the new firm won acclaim for its Lockheed Vega, an advanced, cantilever monoplane with a monocoque fuselage and wooden construction. Having achieved recognition as a talented designer, Northrop left to form his own company, which eventually became a permanent fixture in the aviation business as the Northrop Corporation.

During 1929, Lockheed became the target of the Detroit Aircraft Corporation, one of the aggressive holding companies of the Lindbergh boom and the bullish stock market. Meanwhile, Allan Lockheed also departed, leaving the shell of a once-innovative aeronautical enterprise as the Detroit Aircraft Corporation foundered during the Depression. Lockheed Aircraft itself caught the attention of Robert Gross, a shrewd investment banker. In 1932, forming a partnership with a pair of knowledgeable aircraft builders, he paid a bargain-basement price of $40,000 for the assets of Lockheed Aircraft. Allan Lockheed returned as a consultant; Gross wheeled and dealed to pull in new financing; and a nucleus of bold, young designers (including Clarence "Kelly" Johnson) joined the company. A series of compact, high-performance airliners and speedy fighters in the late 1930s set the stage for remarkable growth during World War II.[3]

In the aftermath of World War I, the war to end all wars, many observers expected to find a lucrative aviation market in the civil sector. But commercial aviation was slow to develop, hampered by the lack of navigational aids, suitable transport designs, and the total absence of uniform safety standards enforced by the government. The evolution of aerial commerce and a market for commercial transports rested heavily on the pioneering routes of the U.S. Air Mail Service, launched in 1918. Airmail service remained a federal monopoly until 1925, establishing a viable service and collecting a clientele. Early experiences with airmail routes over long distances were hardly satisfactory, since mail planes could only fly during daylight hours. Overnight mail carried by railroads remained a basic operation for most businesses during the early postwar era. By 1924, the Post Office Department's system of automatic beacons flashed coded route signals to planes flying by night, and the coast-to-coast delivery time came down to 24 hours, or two full days faster than rail delivery. With this kind of service, airmail offered a valuable advantage and came of age. Many businesses also began to trim inventories and warehouse stocks, relying on air freight to meet sudden demands.

Although several private operations had tried to make money from passenger routes, none had really been successful. The Air Mail Act of 1925 marked an opportunity to build new passenger ventures, using federal airmail contracts to provide a steady source of income to back up the passenger trade. Comfort and service remained marginal. Passengers generally had to sign a waiver allowing the carrier to leave them anywhere en route if a more lucrative load of airmail showed up. Travel in noisy, drafty, cramped passenger accommodations did little to develop air travel as an attractive alternative to trains or busses. Eventually modifications to airmail contracts caused fledgling airlines to order new designs built to carry a number of passengers in relative comfort. Some designers had already looked ahead to larger passenger airliners, so that planes such as the redoubtable Ford Tri-Motor were quickly pressed into service. With seats for 12 passengers and a cruising speed of 100 MPH, the Fords became the proud flagships of many early airline operations.

Through the 1920s and early 1930s, the Fords and similar types from Boeing served as the advance guard in establishing routes, building a passenger clientele, and providing airlines with the kinds of operational experience that allowed them to make the most of truly modern equipment soon to come. More cities added airline terminals, and the flying public began to expand. Movie stars and public figures became regular patrons, adding glamour and a cosmopolitan flavor to air travel. More importantly, larger numbers of business people took to the airways as a means of saving time and maximizing their effectiveness. The airlines in turn recognized the need to pay attention to marketing, so that standards of service and travel comfort improved in order to retain passenger loyalty. Still, production remained low, with only a few dozen planes of any type delivered to buyers.

In terms of quantity, the largest number of airplane orders came from sources that represented what was known in the prewar years as personal or private flying, and defined as general aviation after World War II. Many such "light planes" in the 1920s carried only the pilot and a passenger, and most operations involved training and pleasure flights in compact, open-cockpit designs. For several years following the Armistice of 1918, a huge surplus of Curtiss JN-4 "Flying Jenny" trainers and similar military planes remained available at rock-bottom prices, discouraging sales by private manufacturers and hampering the evolution of new designs. On the other hand, the supply of cheap, wartime surplus provided affordable airplanes for a growing variety of postwar fliers who began putting them to new, practical uses.

During the 1920s, light airplanes found expanding duties in aerial photography, surveying, crop treatment, forest fire patrol, air ambulance service, and air express. Their fastest growing market involved business flying. The expansion of business activities during the 1920s put a growing premium on the effective use of time. Many business negotiations demanded more than telegraphic messages and phone calls; they required physical presence of the principals. Paved, all-weather roads were still rare in many parts of the nation; rail travel required time; and the airline industry did not offer convenient schedules to many cities. Business people more frequently turned to a new generation of single-engine, light airplanes with enclosed cabins for four people, flown by the business owners themselves. Some corporations acquired larger aircraft including the 12-passenger, trimotored planes used by the airlines. Still, the vast majority of light planes came from small manufacturers: Beech, Cessna, Fairchild, Stearman, and many more whose fortunes rose and fell during the postwar decades.[4]

During the 1920s, Waco was one of many companies to build fabric-covered biplanes with open cockpits. Hundreds of planes like this one formed the basis for what was later known as the light plane or general aviation industry and helped build the market for components and replacement parts, an important aspect of the aviation industry's infrastructure. *Courtesy of the National Air and Space Museum.*

Slowly, the role of aviation in the U.S. Army also became more visible, beginning a long trek toward autonomy, including an upgrade to Army Air Service in 1920. Creation of an air war college in 1920, permanently located at Maxwell Field in Montgomery, Alabama, in 1931, signaled a new level of sophistication in doctrinal theory; this development also marked the start of logistical planning. At McCook Field near Dayton, Ohio, the army conducted flight research and engineering studies that contributed to the development of supercharged engines, improved fuels, structural improvements, and other advances. All this translated into a postwar generation of trainers, fighters, attack planes, and bombers, although the Army's fascination with the doctrine of strategic bombing led to deficiencies in other types of combat planes on the eve of World War II.[5]

At the start of the 1920s, manufacturers continued to build planes from wood and relied on fabric to cover the fuselage, tail, and wings. Biplane designs remained in frontline service into the early 1930s, although Curtiss P-6E Hawks used welded steel tubing for the framework of the fuselage. Wings still used wooden spars and ribs, and the plane was covered by doped fabric with metal panels used around the engine cowling. By 1934, the Boeing P-26 began operational service, featuring monoplane design with a cantilever wing and all-metal construction. Still, the pilot continued to occupy an open cockpit and the landing gear remained fixed; bombers also appeared with open cockpits. Some of the photos of the era suggested skies filled with these kinds of planes. In reality, production remained low in comparison with World War I: production of the P-6E totaled just 45 planes; the P-26 reached 134, in addition to three prototypes and 12 built for export.[6]

Similarly, the U. S. Navy air arm experienced evolutionary changes during the 1920s and early 1930s, when the introduction of fast, modern aircraft carriers like the *Lexington* and *Saratoga* set the stage for revolutionary changes in naval warfare. Improved designs appeared for patrol planes, torpedo bombers, fighters, and other aircraft. The Navy's Bureau of Aeronautics encouraged greater use of metal components in naval aircraft and gave special attention to the development of anticorrosive metals and alloys. The U.S. Navy cooperated with the NACA in promoting development of more powerful radial engines, whose reliability and simpler maintenance made them desirable for long missions over the open ocean and for repair within the cramped confines of carrier decks. Production runs for U.S. Navy and Marine fighters also remained modest during the era and consisted primarily of sturdy biplanes. The Boeing F4B-4, which made its first flight in 1932, featured an all-metal fuselage but retained wooden wing structures covered with fabric, and totaled just 92 deliveries. The Grumman F3F-2, ordered by the Navy in 1936, used metal construction and added retractable gear along with an enclosed cockpit. The U.S. Navy's last biplane fighter, production of all F3F models totaled 164 airplanes. All of these Army and Navy fighters had top speeds ranging from about 190 MPH for the Curtiss Hawk to 260 MPH for the Grumman F3F series.[7]

The steady demand for military planes, civil aircraft, and commercial transports meant that engine manufacturers turned out thousands of engines during the 1920s and early 1930s; production and sales of spark plugs, fuel pumps, instruments, tires, rivets, and a variety of miscellaneous parts represented a lively business. The majority of planes and engines were sold domestically, although exports began to increase especially in the case of American engines, which were incorporated into a number of aircraft manufactured by foreign designers.[8]

Aviation was still not big business compared to U.S. rail and auto enterprises, but the outlines of an essential infrastructure became evident for the first time. The aviation industry had begun to take on substance.

Trends in Manufacturing

After the war, Grover Loening played a central role in blocking a British move to sell thousands of battered war-surplus planes and motors in the United States. Loening revealed many of the items to be unsafe and argued convincingly that the domestic American industry would suffer. Continuing to champion the cause of American airplane designers, and suspicious of the Manufacturers' Aircraft Association as a foil for the auto producers, Loening joined Lester Gardner (editor of *Aviation*) and others to organize the Aeronautical Chamber of Commerce of America (ACCA) in 1919 as a specialized trade association and became its president in 1922. Formation of the chamber represented one of the significant institutional hallmarks of the fledgling industry, providing a source of relevant statistics for manufacturers, investors, banks, and federal agencies.

The founders of the ACCA hoped to win the endorsement of major manufacturers and encourage greater production of commercial planes, thereby reducing the industry's reliance on highly specialized military types, but success was slow to come. Major concerns like Boeing, Martin, Douglas, and others who avoided membership preferred to rely on their own congressional contacts for political interchange and continued to find military contracts for high-performance designs more appealing than airline equipment. More to the point, military orders generated more income, even though civil production grew rapidly after the mid-1920s. In 1928, the total value of some 3,000 civil aircraft, worth $17 million, was still $2 million less than the value of 1,219 military planes built by only a few contractors.

The figures in Table 2.1 illustrate the significance of continuing military demand 1927–33; the civil sector gained headway, but production remained hampered by the availability of cheap military surplus aircraft. The Air Mail Act of 1925 became a notable stimulus for civil designs, followed by sales triggered after Lindbergh's flight, a factor that also encouraged stronger support for military procurement. The Depression brought declines, although the resilience of the civil sector was encouraging. It must be remembered, however, that less costly light planes accounted for the bulk of production in this sector; the number of planes in scheduled air transport service during the mid-1930s came to only about 300 aircraft.

TABLE 2.1.

Military and Commercial Business of Eleven Major Aircraft Manufacturers, 1927–1933 (in millions of dollars)

	Military Aircraft Sales[1]	Commercial Aircraft Sales[2]	Military Sales as Percent of Total Sales
Douglas	$14.42	1.41	91
Boeing	10.32	7.03	59
Martin	9.88	0	100
Curtiss	7.27	2.60	74
Chance Vought	6.46	2.18	75
Keystone	5.95	1.77	77
Consolidated	4.29	1.11	79
Great Lakes	2.44	.90	73
Grumman	.44	.15	75
Subtotal	**61.47**	**17.15**	**78**
Wright	30.57[3]	22.43	58
Pratt & Whitney	33.37[3]	18.83	64
Total[4]	**125.41**	**58.41**	**68**

1. Army and Navy production and design contracts
2. Includes exports, most of which were military
3. Engines
4. Due to depreciation, intangible assets, and rounding, the sum of the individual items cited does not equal the total.

SOURCE: Jacob A. Vander Meulen, *The Politics of Aircraft: Building an American Military Industry* (Lawrence: University Press of Kansas, 1991).

During the late 1920s and early 1930s, the leading aviation manufacturers (see Table 2.1) continued to derive their income from military sales, and this factor dominated their thinking about which market to follow.[9]

While the aviation business remained small in comparison to other industries, it began to emerge as a resilient new enterprise with a comprehensive infrastructure. In 1929 Victor Selden Clark, author of an authoritative survey of American industry, declared that the manufacture of aircraft occupied a position of "minor importance" in the hierarchy of American enterprise. He acknowledged that the value of aircraft manufactures had doubled from $7 million to $14 million between 1925 and 1927. Such growth had promise for the future, Clark said, but in his opinion the main value of aircraft manufacture to the engineering industry was in terms of innovations in engineering technique, such as lightness and close tolerances.

In 1929, after the Air Mail Act and Lindbergh's transatlantic flight, the industry reported a total value of $70,334,107 for all aircraft, parts, and equipment. As an increasingly accepted mode of modern transportation, aviation was assisted by this momentum to weather the troubled economic climate of the 1930s. In spite of the Depression, the book value of the three main sectors of the transportation industry indicated a steady increase by aviation. The following figures demonstrate the dominant position of the automotive industry, but show that the aircraft industry experienced a significant rate of growth, even in the depressed thirties.[10]

$ (in millions)	*1919*	*1929*	*1937*
Total book value	$2,478,000	$3,476,000	$3,673,000
Motor vehicles	$1,936,000	$2,742,000	$2,792,000
Locomotive and railroad equipment	$523,000	$616,000	$680,000
Airplanes	$19,000	$118,000	$201,000

The new aviation industry was strongly rooted in a sound structure of modern technology, resting on a foundation of conventional although comparatively modern enterprises for parts and supplies such as B.F. Goodrich, Standard Oil, AC Spark Plug, and Westinghouse. These contemporary giants exhibited at aircraft shows and their presence lent a sense of familiarity and reliability to the vehicles of the future. The enterprise of aeronautics was further buttressed by a growing number of specialized aviation subcontractors and suppliers of parts and accessories. The ranks of subcontractors and vendors swelled the aviation business during World War I. Caught in the early postwar doldrums, many were able to persist, and a review of the trade journals during the twenties shows their steady growth throughout the decade. Whatever commercial mode they followed, such businesses represented another important factor in the aeronautical infrastructure—unmistakable evidence of aeronautics as a mature enterprise able to attract a broad spectrum of vendors.

Many specialized businesses floundered or became absorbed by larger competitors over the next few decades; a few hardy pioneers survived into the 1990s. The Hartzell company, of Piqua, Ohio, began as a sawmill in the 1870s. Its specialization in oak and walnut products brought early inquiries from the Wright airplane company in Dayton, where members of the Wright and Hartzell families lived half a block from each other. Hartzell supplied wooden propellers for Curtiss "Jenny" trainers during World War I and became a major supplier of propellers for light planes and other aircraft through World War II. The postwar era brought metal propellers and continuing success. Another business was the EDO Corporation, formed in 1925, whose company name derived from the initials of its founder Earl Dodge Osborn. EDO did component fabrication and became world famous for the floats used to replace wheels on conventional airplanes, turning them into floatplanes. As a diversified supplier, it was still going strong in the 1990s. In 1923, Osborn had helped Lester Gardner expand *Aviation* magazine; he became its publisher until selling out to McGraw-Hill in 1929. The publication evolved as *Aviation Week and Space Technology*. The Irving Parachute Company, founded in 1919, was an industry pioneer that persisted through the 1920s and 1930s, altering its name after World War II to Irvin Industries Incorporated.

One of the unique devices of the era came from an aviation enthusiast named Edwin Link. During the late 1920s, Link borrowed some bellows, push-rods, and linkages from his father's pipe organ company to construct a mechanical flight simulator. The device's ability to imitate the motion of an aircraft in flight marked a distinct advantage over static cockpit training equipment. For several years, the only customers were amusement parks, until military orders in the 1930s began to multiply. Link simulators became an essential factor in training thousands of pilots during World War II. Into the 1990s, Link's corporate descendants (and several competitors) became indispensable for training combat fliers, airline pilots, and astronauts.[11] Another pioneer figure, Igor Sikorsky, was one of a number of Russian émigrés with aviation backgrounds who fled to America after the Bolshevik revolution in 1918. Sikorsky formed his own company in 1923, Sikorsky Aircraft, with funding from a group of investors including Russian compatriots like the famed composer Sergei Rachmaninoff. The company pioneered a series of large flying boats, but Sikorsky also pursued a lifelong fascination with helicopters. In 1939, he piloted the first practical helicopter, the Sikorsky VS-300, on a successful flight. In the meantime, Sikorsky had become a division of an aggressive venture known as United Aircraft.

The aviation industry, like so many others in the twenties, experienced considerable consolidation through acquisition and merger. The trend was evident in the evolution of airline companies as well as in manufacturing. In most instances, the operators remained separate from the manufacturers, although several extensive combines emerged. One of the largest, and most broadly based, was United Aircraft and Transport Corporation, organized early in 1929. In addition to Boeing Air Transport, Inc., and the Boeing Airplane Company, the United combine swallowed up Pratt & Whitney, a leading power plant manufacturer, along with two propeller manufacturers renamed Hamilton Standard Propeller Corporation. United's interest in the private plane sector was represented by the acquisition of Stearman, in Wichita, Kansas. By the end of the year of its founding, United's various properties included four other airlines and three other aircraft companies, including those of Chance Vought (naval aircraft) and Igor Sikorsky (large planes and flying boats). United Aircraft emerged as an integrated aerial octopus, building a complete range of civil and military aircraft, running its own airlines, and supplying itself with engines, accessories, and miscellaneous services.[12]

The experience of Grover Loening's company reflects that of many smaller aircraft firms during the period of mergers. During the early 1920s, the Loening company developed several different planes with indifferent commercial success, and Loening was required to keep the company alive by building improved wings for the dozens of DH-4 planes operated by the Post Office and the military services. About 1923, he finally got the Army Air Corps interested in his new amphibian design, and subsequent orders for its patented integration of engine, hull layout, and retractable gear made Loening a profitable operation. With continuous production, the Loening company of the 1920s served as the training ground for many young aviation professionals, including Leroy Grumman. Fed by the excitement over aviation stocks after Lindbergh's transatlantic flight of 1927, a wave of mergers inexorably transformed the list of smaller manufacturers. Loening Aeronautical began to worry about competing with the giants that were taking shape, even though the company was impressively solvent. But the trend of events argued for selling out, and Loening struck a profitable deal through friends in a banking firm that already owned Keystone Aircraft of Bristol, Pennsylvania. In 1928, Loening sold Loening Aeronautical to a group that eventually formed the Curtiss-Wright Corporation in 1929.[13]

The lot of military manufacturers in the 1920s and much of the 1930s was never an easy one. Government requirements often hampered innovation and research, retarding the capability of American products relative to certain foreign developments. One company might expend costly time and effort to develop a prototype with outstanding performance, but Congress insisted on awarding production contracts to the lowest bidder. A second company's low bid invariably walked (or flew) away with the prize of contracts, leaving the original innovator with a silent factory, idle workforce, and loss of proprietary technology. Contract winners had to face additional hurdles; they had to post performance bonds, and the practice of advance payments was illegal. Manufacturers depended heavily on their investors to stay afloat for years while production contracts were completed. Fortunately, the wood and fabric construction typical of aircraft during the 1920s meant that preproduction costs were manageable, and the comparatively low production runs permitted the use of equally modest manufacturing halls. Obviously, the fabrication of flying machines had to be done patiently and carefully, but traditional craft skills of woodworkers and carpenters remained available, keeping labor costs stable.

Still, labor represented the largest expense in aircraft production, running at 50 percent of value added by manufacture. Wages varied across the country but were not notably higher than in other industries. Aviation workers "lacked effective unions and the power to prevent their employers from forcing upon them the costs of their industry's labor intensity and of Congress's maintenance of artificial competition through nonrecognition of design rights," one historian observed later. "With few ways to utilize cost-cutting machinery, manufacturers had few alternatives but to impose upon their workers much of the costs of price competition."

The labor force included a number of women, some of whom helped assemble finished components such as wing ribs, and many more who stitched and attached fabric to wing panels, fuselage sections, and tail surfaces. The inability of men and women alike to combat the reduction of wages and job losses became increasingly evident in the years immediately after the crash of 1929, when the industry slashed wages by as much as 40 percent. At Curtiss-Wright, a firm having an admirable tradition of welfare capitalism, including cafeteria, sports teams, and an employee system for loans, pensions, and other services, a bitter strike during 1930–31 followed the company's stringent measures to cut losses of some $2 million and deal with its plant overcapacity of 75 percent. The strikers eventually agreed to severe terms (lower wages, slashed benefits, rigid work rules, and more) and the result not only underscored the antagonisms between labor and management, but also prompted investigations by the War Department.

The Curtiss-Wright affair embodied resentments throughout the aviation community during the depression era. During the 1930s, wages dipped from about $30 per week to $25 by 1935, and rose again to $30 by 1939. These rates appeared to be somewhat better than those of other industrial workers, due to the number of skilled workers required in aviation. The total aviation workforce remained small: 9,600 people in 1933, or about 0.002 percent of manufacturing employment, rising to 64,000 in 1939, or 0.007 percent.

The big mergers did not put all of American aviation in the hands of all-powerful monopolistic interests. Walter Beech, Clyde Cessna, and others kept building planes in Wichita; Donald Douglas on the West Coast and Leroy Grumman on the East Coast continued to preside over independent companies with military contracts that survived the mergers and even the Depression. The stock market crash of 1929 ruined many aviation companies, along with many other Wall Street favorites, but it was the New Deal that delivered the most telling blow to the big aeronautical combines of the late twenties. Following the Air Mail Act of 1934, manufacturing and transportation were permanently divorced, and the New Deal's antipathy to overintegrated business kept them that way. The structure of the American aviation industry continued in this pattern, with basic separation of aircraft, power plant, and transport companies.[14]

Aircraft manufacturing created different types of jobs in traditional centers of manufacturing and brought new opportunities to other areas that were particularly suited to the special requirements of aircraft production. Some companies, such as Boeing in Seattle, were content to stay where they were. Many manufacturers preferred to remain in the East, close to transportation centers and skilled labor for the specialized production of sensitive instruments and aircraft fittings. During the 1920s, New York remained the leading state for aircraft production and employment. The level plains of the western prairies offered fewer topographical hazards plus reasonable weather, attracting some of the pioneer designers and manufacturers who liked the promising flying and testing conditions there. But the availability of financing played a key role. Wichita, Kansas, the locus of a Midwest oil boom, generated a source of risk capital tapped by aggressive aeronautical entrepreneurs. Famous names such as Stearman, Beech, and Cessna became associated with Wichita from the early years of the twenties. The promise of extended periods of flying weather attracted many manufacturers to California, whose climate had other advantages as well. Aircraft construction and assembly called for facilities with plenty of space to accommodate the growing wingspan and fuselage length of modern planes. Such buildings required less heating expense in California, and were subject to less depreciation resulting from attrition due to weather. The availability of risk capital also played a key role in California, where oil money found its way into aviation investments. Additionally, adventurous entrepreneurs from the state's publishing industry, along with various industrialists who saw value in diversification and encouragement of new industries, supported aviation as a promising new technology.[15]

Sources of Engineering Progress

The products manufactured by these early companies evolved from diverse sources. Some firms, like the light plane manufacturers, essentially relied on off-the-shelf concepts and hardware. Some, like Loening, incorporated proprietary ideas into their products. European influences played a significant role, often introduced by émigrés, technical societies, and through the curricula of a growing number of professional college programs. Many manufacturers made strides through military contracts. And all of them began to rely increasingly on studies, reports, and engineering trials carried out by the National Advisory Committee for Aeronautics—the NACA.

Research disseminated by the NACA had a fundamental impact on the success of American aviation technology. Its reports served as the basis for many innovations that were built into American civil and military aircraft. This impact escalated when the NACA moved beyond its originally mandated role as an advisory body and eventually acquired its own laboratory facilities, completed at Langley Field, Virginia, in 1918. Realizing the need for expertise in theoretical aerodynamics, NACA authorities decided in 1920 to bring to America a German expert, Dr. Max Munk, even though it required an executive order from President Woodrow Wilson to cut through the red tape—technically, America was still at war with Germany. NACA's first variable-density tunnel, which incorporated Munk's advice, went into operation in 1923, and yielded basic data for a series of pioneering NACA reports on wing improvements and other factors. Additional facilities were added; a full-size tunnel built in 1927 was used to develop one of the most successful aeronautical innovations of the twenties, the NACA engine cowling.

Most American planes of the postwar decade mounted air-cooled radial engines, with the cylinders left exposed to the airstream in order to enhance cooling. NACA's tunnel tests revealed that this style of installation accounted for as much as one-third of a plane's total drag. Although significant work on cowled engines proceeded elsewhere, particularly in Great Britain, NACA work provided the most dramatic success. After hundreds of tests, a NACA technical note by Fred E. Weick in November 1928 detailed a cowling design that enclosed the engine in such a way that cooling was enhanced while drag was sharply reduced at the same time. The innovation yielded dramatic results. During a transcontinental test flight, a 157-MPH plane equipped with the new cowling averaged 177 MPH. In terms of its effect on air transport, E.P. Warner wrote, the introduction of the NACA cowling was "staggering." By 1932, according to one estimate, the operational efficiency of cowled American aircraft represented a savings of $5 million. Without detracting from the NACA's careful engineering development of a notable contribution to aeronautical advance, its work on the cowling probably involved somewhat less originality than NACA partisans claimed. The NACA itself never took out a patent on the cowling, perhaps reluctant to stir up trouble in British circles, where Hubert Townend's earlier development of the "Townend ring" represented an important step forward. As one prominent NACA engineer, H.J.E. Reid, wrote to headquarters in 1931, "It is regrettable that the [Langley] Laboratory, in its report on cowlings, did not mention the work of Townend and give him credit."

At its formation, the NACA embodied certain progressive goals in terms of being a pool of skilled advisers, and the agency's coordinating functions prevented duplication and waste. Moreover, the NACA was to be immune from the machinations of business and industry. By 1930, the departure of forceful, research-oriented personalities such as Max Munk resulted in a more conservative style within the NACA, with increasing emphasis on practical and applied research, often for manufacturers, as opposed to long-term theoretical projects. Nevertheless, the aeronautical pioneering and the dissemination of information by the NACA ranked at the top among the most important factors of American aviation progress in the 1920s and 1930s.[16]

The progress of aviation in various technical areas was enhanced by a growing interest in education for aeronautical engineering. Although the Armour Institute of Technology in Chicago reportedly offered aeronautical courses as early as 1910, one of the best known of these pioneer programs was initiated by Jerome C. Hunsaker, who became interested in the field of fluid dynamics through his studies in naval architecture at Massachusetts Institute of Technology (MIT). Hunsaker himself translated a book on aerodynamics by Alexandre Eiffel, the well-known French engineer responsible for the Eiffel Tower, which was used in the first course in aeronautics at MIT in 1913—taught by Hunsaker. One early student was Alexander Klemin, an English emigrant who became director of the aeronautics program when Hunsaker left MIT in 1916 to take charge of the Navy Aircraft Division in Washington. Klemin later helped establish the influential magazine *Aviation* and did wartime research at McCook Field. During World War I, the general course in aeronautics at MIT evolved into a strict discipline of aeronautical engineering.

Felix W. Pawlowski, a professor in mechanical engineering at the University of Michigan, also developed pioneering courses in aeronautical studies during 1913. Originally educated as an engineer in Poland and France, his courses featured lectures by visiting European experts and introduced a continuing thread of European research into the curriculum. Pawlowski's first candidate for the bachelor of science degree in aeronautical engineering received his diploma in 1917, and postgraduate work was offered a few years later.

During the 1920s, degree programs evolved across the nation, often as the result of creative funding from the Daniel Guggenheim

Foundation for the Promotion of Aeronautics. With the support of this outstanding private philanthropy, and with growing interest in state colleges and universities, the pool of trained aviation professionals became significant. By 1929, a survey by one aviation magazine reported a total of 1,400 aeronautical engineering students enrolled in 14 colleges and universities across the United States. With additional help from the Guggenheim Foundation, Cal Tech was able to attract the internationally acclaimed aerodynamicist Theodore von Karman to direct its Guggenheim Aeronautical Laboratory. Von Karman accepted Cal Tech's lucrative offer in 1929 and arrived in the United States in the spring of 1930. His leadership at the university sparked a new era in the field of theoretical aerodynamics and helped the United States educate a new generation of both educators and practitioners. Nationwide, the growing pool of trained aeronautical engineers played an important role in assisting in the design of planes that elevated aviation and air travel to a reliable, economical, and mature technology.[17]

Elements of Modern Design

As significant as NACA's aerodynamic work had been to the development of aeronautics, the progress of American aircraft design in the twenties hinged on the cumulative effect of several additional technological trends. Attempts to achieve streamlining in order to reduce drag began before World War I, although the most successful designs were the Junkers, Fokker, and Dornier aircraft with cantilevered wings that were produced in Europe after the war. In Germany, Junkers and Dornier also pushed ahead with metal wing and fuselage structures. Early in the postwar era, the NACA had installed a technical representative at the American embassy in Paris in order to keep informed on the latest foreign developments. The all-metal cantilevered monoplanes attracted the attention of the NACA early in the twenties and influenced further examination and exploitation of such features in America. One important result was the Ford Tri-Motor of 1926, with its metal structure and cantilevered wing. Other research involved stressed-skin construction, using the aircraft skin itself to carry more of the loads imposed on the aircraft in flight. This approach eliminated many internal trusses and braces within airframe design. In Germany, Adolph Rohrbach and his collaborator Herbert Wagner represented the vanguard in this research, which became widely known in the United States after Rohrbach published a paper on the subject in the *Journal of the American Society of Automotive Engineers* in January 1927. Streamlining and the stressed-skin technique were evident in the Lockheed Vega of 1927, designed by John K. Northrop, who had independently devised his approach apart from German influence. But the Vega featured wooden and molded plywood construction. Metal construction for fuselage and wing structures heavily influenced subsequent United States airliner designs of the late twenties and early thirties.[18]

The radial engine at first posed something of a problem in terms of aerodynamic drag, but the evolution of this type of power plant was fundamental to dependable aircraft operations. While most European engine manufacturers favored the more streamlined liquid-cooled power plants after World War I, the U.S. Navy evinced strong interest in the radial because of its easier maintenance and power-to-weight ratio for operations from the limited confines of carrier decks. The moving force behind the Navy's interest in radials was Captain Bruce Leighton, who supervised engine procurement for the Bureau of Aeronautics. Charles L. Lawrance had developed a very attractive radial engine design in 1921, but his Lawrance Aero Engine Company lacked the capacity for volume production. Leighton literally maneuvered the Wright Aeronautical Corporation into the purchase of Lawrance's company, which led to the very successful series of Wright Whirlwind power plants.

The Wright-Lawrance merger, however, led to some internal problems, and a dissident group of engineers decamped in 1925 to set up a new company, leading to the organization of Pratt & Whitney. In just six months, the new company developed the 425–hp Wasp, incorporating a number of important technical advances, including improved cylinder heads based on the work of S.D. Heron, then working for the Army at McCook Field. Heron, one of the most important technical innovators of the era, had come to the United States in 1918 from England, where he had done considerable experimentation on radial engine designs. Heron joined Wright Aeronautical in 1926, where he had an important role in designing the Wright J-5 engine, the classic power plant that powered Lindbergh's *The Spirit of St. Louis*. Together with Pratt & Whitney, Wright Aeronautical continued to introduce refined radial power plants that gave American aircraft an enviable record of efficiency and reliability, contributing to United States leadership in world aviation.[19]

As engine performance curved upward in the twenties, the quality of aviation gasoline assumed greater significance. During the early years of aircraft development from 1900 to about 1918, recalled one chemical engineer, "almost anything was used in an airplane engine that would enable it to operate." By 1921, fuel research for automotive use concentrated on metallorganic compounds, resulting in a formula based on tetraethyl lead. Additional experimental work led to a practical production method for high octane gas, and the Ethyl Gasoline Corporation (formed by General Motors and Standard Oil of New Jersey) was established in 1924. Within three years, both the Navy and Army began using the additive, although commercial airline companies held back extensive use for several years because of the additional cost involved. By the close of the decade, the major petroleum companies had built new production facilities for aviation gas, and airlines took advantage of lower prices for high octane fuel.

All of these trends in research, development, and infrastructure pointed toward the aircraft designs of the 1930s. Other notable activity of the era included development of controllable pitch propellers, wing flaps, navigation, radios, flight simulators, and additional systems.[20] Manufacturers like Boeing inaugurated a significantly different generation of air transports that drew on several advanced design concepts, and spent much more effort in building a plane with passenger comfort in mind. Collectively, Boeing's agenda acknowledged the requirements of airline operators for an efficient, economical transport that would appeal to its customers. In short, the manufacturer began producing for the market.

The Boeing Model 247

Charles Lindbergh's conquest of the Atlantic in 1927 buttressed America's belief in the superiority of its aviation achievements. But European pilots soon duplicated Lindbergh's feat, and European air force equipment seemed superior. Boeing's introduction of its Model 247 represented a new level of competition. The evolution of the Model 247 not only bore evidence to the role of international contributions and U.S. market influences, but also revealed the problems of manufacturers and designers in producing new transports that catered to consumers while making money for the operators at the same time. The gestation of the 247 also demonstrated the problems inherent in perfecting advanced tooling, training workers, and assuring that suppliers met specifications while keeping deliveries on schedule.

In the wake of Lindbergh's flight, any lingering doubts about the audacity and operational expertise of Europe's aviators and the technical quality of its planes and power plants were knocked into a cocked hat by the aerial extravaganza mounted by the Italians in 1933. The "Century of Progress International Exposition," held in Chicago during 1933–34 (often referred to as the Chicago World's Fair of 1933) offered a unique occasion to showcase Italy's prowess in aviation. Italy's flamboyant minister for air, General Italo Balbo, had led a group of Savoia-Marchetti

SM-55X twin-hulled flying boats across the South Atlantic in 1930. His new expedition included a mass formation of no less than 24 of the same planes, proceeding from Rome to Amsterdam, Iceland, and Canada before settling down on Lake Michigan—at the portal of the Century of Progress itself. Departing Rome on July 1, 1933, the Italians arrived over Chicago 15 days later, holding the tight formation enforced by Balbo. After 10 days of heady reception and raves from the press, the disciplined Italians took off for the return trip to Rome, recrossing the Atlantic in another historic aerial feat. Only two planes were lost after capsizing in landing and takeoff miscues.

The effect of Balbo's massed flight—a tour de force of airmanship, planning, and execution—was electrifying. The overall reliability of planes and engines over such great distances carried a forceful message about the potential of European aviation and airpower. There were political overtones to the flight as well, as it marked the tenth anniversary of Mussolini's rise to power. Writing for *Aero Digest*, World War I hero and postwar airline executive Eddie Rickenbacker minced no words. "The advent of the Italians will render us a real service if it jogs our national consciousness into the realization that we have now lagged in the air until we are now fourth in terms of air strength," he said. Actually, help was on the way. The Boeing 247 that went on display at the Chicago World's Fair represented the cutting edge of modern technology as it was poised for adaptation into a new generation of military and civil aircraft.[21]

After some problems, Boeing's airplane manufacturing operation stabilized in the 1920s, based on orders for military designs. In 1926, when a friend proposed that the company should become involved in the airmail contract business, Boeing agreed. Boeing Air Transport won a contract in 1927 and went into business with the Boeing Model 40-A, an open-cockpit biplane with a good performance for its day. The company delivered 72 models of this type. The airline acquired other routes and the manufacturing division fielded successively better airplanes. The Model 80 series mirrored the period's preference for trimotor designs in building larger passenger planes; a twin engine plane could not fly if one engine failed. Carrying 12 to 18 passengers, the 80 series used steel and aluminum tubing in its primary structure, even though its covering was mostly fabric. Introduced in 1928, only 16 were built, costing about $75,000 apiece.

At roughly the same time, an amendment to the original Air Mail Act of 1925 set the stage for a series of significant changes. Under the new legislation, known as the McNary-Watres Act, the system paid carriers a rate calculated at one cubic foot per mile even if space on the plane remained empty. The government wanted to encourage airlines to carry passengers, and it did.²² The government's policy created a market for a new generation of larger, multi-engine airliners that led to the evolution of the Model 247 airliner of the 1930s.

Jack Northrop, creator of the remarkable Vega, had felt compelled to progress beyond wooden construction into metal aircraft. In 1928, he and Lockheed parted, and Northrop launched his own company, Avion Corporation, to build a new series of metal planes, beginning with (appropriately) the Alpha. With its cantilevered, stressed-skin wing and metal fuselage of monocoque construction, the Alpha had captured Bill Boeing's imagination. As chairman of the board for United Aircraft, Boeing's endorsement cleared the way for Northrop's fledgling company to become part of the United Aircraft conglomerate. The minutes of United's Technical Advisory Committee of the period bear testimony to Northrop's influence on the group's subsequent aircraft designs, including those of Boeing.²³ Boeing's first step in this new direction was a single-engine design, the Model 200 Monomail.

In the meantime, work at NACA's Langley laboratory resulted in a variety of designs for engine cowlings and a highly effective scheme for mounting engines in the leading edge of a wing, setting the stage for successful twin-engine transports. Boeing studied these NACA reports, melded them with design experience from its Monomail experience, and brought out the notable B-9 bomber of 1931. Structural refinements learned from the Monomail, including semi-retractable gear and metal construction, gave the twin-engine B-9 a cruise speed of 165 MPH and maximum speed of 188 MPH, faster than fighters in service at the time.²⁴ Even though an Air Corps production contract went to the rival Martin B-10 bomber, Boeing now emerged as an aeronautical contender.

There is little doubt that the 247 drew heavily on Boeing's work on the B-9 bomber. The plane incorporated many technical features and structural details developed for the B-9 by the company in cooperation with the Materiel Division of the Air Corps. During 1932, when the Japanese expressed interest in the 247, the Air Corps became alarmed. Given Japan's expansion into Manchuria and China, the U.S. government feared that acquisition of the 247's technology would advance Japanese military designs. In confidential correspondence, Boeing officials expressly reassured the Air Corps that none of the company's advanced airliners would be sold abroad unless the government approved.²⁵

The design of the 247 was forged in a clash of wills involving partisans of big aircraft opposed by advocates of smaller designs. The debate engaged Boeing personnel as well as people from the airlines who would buy the airplane. At United Aircraft and Technology Corporation (UATC) these issues were thrashed out at high levels in the meetings of the Technical Advisory Committee. George Mead, a power plant expert from Pratt & Whitney, wanted a big plane. So did Igor Sikorsky, who had been building big planes since his early career in prerevolutionary Russia. They agreed that a big plane would offer the sort of seating, lavatories, and service that transcontinental railway passengers enjoyed, enabling the airlines to compete more effectively with train travel.

On the other side was Boeing's chief engineer, C.N. Montieth, who agreed that passenger comfort should be a goal, but argued that anything larger than the planes in current production would create serious problems. Bigger planes would require larger, costlier hangars for maintenance and repair, a negative aspect for struggling airline companies. Pilots, he argued, liked smaller planes because they were more maneuverable in nasty weather: "The minute they get into the least bit of thick weather with the big ships they set them down." Airline officials sided with Montieth for smaller planes but cited other reasons. A representative of National Air Transport argued that several smaller planes could be operated more profitably than one big ship. Moreover, smaller designs would be faster, and the Post Office Department was especially interested in high speed. Eventually, over Mead's misgivings, the Technical Committee opted for a smaller design. By this time, in 1930, the reality of the depression seemed to make a smaller, less expensive design more practical.

The design of the new airliner progressed through several phases. At one point, the favored configuration featured a smallish, twin-engine plane for eight passengers. It had no lavatories and only one pilot, to keep the gross weight under control, and was estimated to sell for about $40,000.

The smaller, high-performance plane received consensus support, with emphasis on performance at the expense of passenger comfort. A mock-up began to take shape.[26] Second thoughts by some airline executives and by Erik Nelson, Boeing's sales manager, resulted in some cautious revisions. A supplemental work order went out for the construction of a new mock-up to have eight seats with a pair of extra jump seats, a passenger cabin having headroom of six feet, and a cockpit for both pilot and co-pilot. Hopefully, the added headroom would reduce the tendency of passengers to trip over the wing spar that jutted up across the cabin aisle. Inevitably, bigger engines entered the picture.

As the design neared finalization, production engineers could pin down the price more accurately. Boeing had wanted to offer the plane for $35,000, but the latest version with two pilots and other changes was estimated at precisely $41,111.25, based on a total order for 60 planes produced at the rate of four per month. The price included a 5 percent handling charge and assumed a profit of 15 percent. By January 1932, rising costs for parts, engineering, and labor put the price at $52,700. After some vigorous protests from United, its president, Philip Johnson, finally realized that the cost was not unreasonable. United's order for 60 planes, for $3.5 million, became official in March 1932. The first production aircraft was scheduled to rollout in September.

Production engineers soon realized that the date was much too optimistic. Management resorted to overtime and decided to rely on subcontract work more than originally planned. The steel inboard spar came from Metallurgical Laboratories in Philadelphia. Chance Vought, a United Aircraft stablemate, fabricated control columns, tail wheel assembly, and miscellaneous components. Among other things, Pratt & Whitney produced the entire retractable landing gear. But Gardner Carr, the plant manager, insisted that major sheet metal components and all other major fuselage and wing sections be manufactured by Boeing itself.

Since the cabin lining added an additional 200 pounds to the plane, considerable effort went into redesign and engineering to save weight. An internal electric starter, listed at 34 pounds, came out. Much thought went into the possibility of lightweight alloys in noncritical areas. Magnesium became the candidate for use in certain bulkheads, cockpit fairing, floors, seats, and instrument panel. Other magnesium uses were debated: hatches, doors, and cowling; mudguards, boxes, brackets, and so on. Magnesium needed additional treatment to inhibit corrosion, and it was quite expensive, but Boeing found it was necessary to use in many areas because 97 pounds more could be shaved from the structure. The designers finally agreed to eliminate a coat of paint for the fuselage, since a grayish-green anodized finish for the Duralumin saved precious pounds. The lavatory compartment remained, but the 247 came with no running water, and parched passengers had to make do with the contents of a two-quart vacuum bottle affixed to a lavatory bulkhead. There was an argument about the lavatory mirror. Philip Johnson, president of Boeing Air Transport (and Boeing Airplane), thought a mirror should be included; Erik Nelson, the sales manager, remained adamant. Women carried personal mirrors in their handbags, he argued, and men did not need it. For the time being, the mirror stayed out.

But the production schedule was now badly skewed and costs continued to climb. Deliveries were reset for December 1932. A representative from National Air Transport, though still part of the corporate family, pointedly asked for a realistic date of assembly on the first planes. Boeing's response remained bureaucratically evasive. "Whatever we might give you now will undoubtedly have to be verified at a later date," came the reply, "so that instead of definitely committing ourselves at this time, we feel it would be more practical to keep your request in mind and inform you of the assembly when we have definite assurance that the date given you will be closely approximate."

Significant problems of a fundamental nature came to light. Like the B-9 bomber, the original version of the 247 used a uniform cross section, a convenient engineering feature that heavily influenced cost estimates and delivery schedules. But the final version included a cockpit for two pilots, not one, so that a wider cockpit necessitated more metal and new cross section designs to blend the cockpit into the fuselage. Additional design changes, enlarging the fuselage itself, forced the wholesale discard of wood dies for fabricating the more complex bulkheads. The steel dies required were not only more expensive, but also more complex to construct and to use in the actual manufacturing process. Under the intense drive to perfect various tooling, workmanship began to disintegrate. As the delivery for the first plane slipped to January of 1933, Boeing decreed that "all personnel working on the first transport are to go to 12-hour shifts except where three shifts can be employed."[27]

In order to make needed deliveries by April, Boeing had to increase its payroll to 2,200 personnel, who worked in three shifts. The influx of untrained workers created problems in the production schedules. Expenses went up again. Boeing's plant manager, Garner Carr, had forecast a unit price of $53,000 per plane. However, for the first 10 units, the cost had escalated to $77,000 apiece. Johnson had once aimed for a target of $45,000 per plane, and the final costs per plane for the production run of 60 units came to $68,000. On that basis Boeing calculated it would probably break even on the 247.

When the inaugural rollout of the Model 247 finally occurred on an April Sunday in 1933, an estimated 15,000 visitors—eager to see the free, six-hour air show and the new, much heralded airliner of the future—were in attendance. Its speed and performance alone made it awesome: capable of cruising at 150 MPH, Boeing claimed it to be 50 percent faster than its competition. The plane could gobble up 485 miles before it had to land for refueling, so that transcontinental travel could occur in half the time. Handled by a pilot and co-pilot, the crew included a cabin attendant to see to the needs of the plane's 10 passengers. The passenger cabin featured the latest standards for air travel, such as soundproofing and a thermostatically controlled heating and cooling system, as well as individual air vents and reading lights. There was a lavatory. Safety was assured with modern navigation instruments, an autopilot, and a two-way radio.[28] The plane continued to enjoy a popular reception during the summer of 1933 as hundreds of passengers marveled at its speed and airborne comforts such as current newspapers,

Introduced in the mid-1930s, the Boeing 247 represented a variety of aeronautical technologies that had evolved during the previous decade, such as retractable landing gear and stressed-skin construction. The challenges of designing and producing the plane characterized many new issues facing manufacturers. *Courtesy of the Boeing Company Archives.*

magazines, and cigarettes, along with salads and sandwiches en route. Perhaps 61 million more marveled at this artifact of advanced American technology while it was on display as a feature of the Travel and Transport Building at the Century of Progress Exposition held in Chicago. Its wheels up, the 247 sat poised atop an arrangement of steel pedestals that lifted it high above the floor. A catwalk erected over one wing gave wide-eyed visitors a chance to ogle the impressive array of instruments in the cockpit and to admire the attractive passenger cabin. A powerful-looking Wasp engine occupied a special exhibit area under the fuselage and a small theater ran a continuous film about the plane and the wonders of airline transportation. Other displays and a terse two-way radio conversation between the pilot and ground control added to the drama of the setting. The future had arrived, embodied by the modernistic Boeing 247.[29]

Although anchored to the ground, the new transport exhibited a mute challenge to Balbo's aerial flotilla, comprised of aircraft that clearly reflected outmoded technology. The American aviation industry appeared to be poised on the threshold of a bright and profitable future.

In 1918, the typical airplane appeared as an open-cockpit biplane, constructed of wood and fabric, and equipped with fixed landing gear. Instrumentation was minimal. Ten years later, a pilot of 1928 would have still felt comfortable with most of the planes to be encountered at a typical flying field. The angular, all-metal Ford Tri-Motors looked different perhaps but possessed few new instruments or other features beyond the 1918 machines. Within five more years, however, the aircraft industry had begun deliveries of a generation of planes that clearly represented the modern era. All-metal, stressed-skin construction was commonplace, wings incorporated flaps, the gear retracted, and instruments had proliferated to keep pace with bigger engines, higher speed, navigation systems, and radio equipment. A pilot of 1928 would not have felt at home in the cockpit of an aircraft of the mid-1930s.

The evolution of military aeronautics represented only part of the picture. A major impetus for different industry products came from the federal government's method of airmail payments to commercial operators. The advent of passenger travel prompted the airlines to specify additional aircraft, ordered with an eye on competitive advantage as well as passenger comfort. The new era had arrived, courtesy of, among other things, federal legislation that had intensified the promise of commercial success. Other significant factors included institutional landmarks like the NACA, college curricula, and professional organizations, as well as a significant leavening of European influences.

Federal initiatives played a significant supportive role. Herbert Hoover felt that commercial aeronautics needed support and cooperation from the government as a contributer to national security. As Secretary of Commerce, 1920–28, he became a central figure in numerous initiatives to promote aviation, particularly in the framing of the Air Commerce Act of 1926, designed to promote the stability and growth of aviation. Hoover's corporatist visions were also reflected in the evolution of the Aeronautical Chamber of Commerce. The chamber, despite problems with some individualistic manufacturers, contributed to useful plans for better utilization of the nation's aircraft production capacity on the eve of World War II and throughout the conflict. The corporatist phenomenon survived as a persistent theme in postwar developments.[30] In the vanguard of aviation professionals, men like Edward P. Warner, trained at MIT, moved back and forth in positions as engineers, educators, and bureaucrats.

CHAPTER 3
Aviation and Airpower 1933–1945

DESPITE THE OBVIOUS impact of the Depression on the national economy and the aviation community, the technological momentum achieved in the 1920s and early 1930s continued to influence aeronautical developments. The light plane industry changed dramatically, both in the nature of its designs and in the variegated use of its products. The design of commercial airliners also continued a rapid evolution, represented by the appearance of the classic DC-2 and DC-3 as well as several other notable examples of big, four-engined airliners. These trends were all part of the major equipment changes that also characterized military aviation and helped establish the industrial base that enabled U.S. manufacturers to achieve the striking production records of World War II.

Institutions like the Aeronautical Chamber of Commerce continued to provide a forum for collecting aviation industry statistics and providing congressional liaison. Groups like the ACC's Export Committee testified to the rising significance of exports within the industry's profit structure. Other institutional organizations appeared, such as the Institute of Aeronautical Sciences, organized in 1933 to promote the interests of professional engineers and designers, publish a professional journal, conduct technical conferences, and encourage educational programs for professional degrees at the college and university level. The IAS idea evolved from the professional concerns of people like Jerome Hunsaker, Edward P. Warner, and others who desired a professional focus outside the Society of Automotive Engineers. Like the founders of the NACA, the founders of the IAS looked to Europe for precedent, using the Royal Aeronautical Society as a model for the new American organization.

In addition to these formalized trade associations, an informal gathering of high-level executives and leaders from civil, federal, and industrial arenas evolved. Basically a social group, it also constituted a forum for informal exchanges among the power brokers of the aeronautical community. This was the Wings Club, which originated in New York City not long after Pearl Harbor. Lauren Lyman, one of its early members, recalled his skepticism when he heard the news—New York was already overpopulated with aviation clubs, and the war made any new social organization seem like a waste of time and energy. But it persisted, a revered institution in the life of American aviation. Among the original incorporators were significant figures like Reed Chambers, Eddie Rickenbacker, William Rockefeller, and Juan Trippe. Bankers, federal leaders, Air Force generals, and manufacturers soon gravitated to the Yale Club, the august and refined setting of the early meetings of the Wings Club.[1]

Formally or informally, these and other entities represented an elaboration of the resources responsible for the notable sophistication of the American aviation community during the 1930s. During World War II, such entities grappled with the problems of manufacturing modern aircraft designs, a challenge aggravated by the complexities of defining military aviation doctrine, choosing appropriate aircraft, and finding factories and workers to build planes in quantity.

The Light Plane Industry

The dreams of thousands of would-be fliers faded as the Depression deepened. Compared to a cost of $850 for a Chevrolet sedan, the typical light plane of the era retailed between $1,500 and $2,500, and maintenance costs on a Piper Cub flown 150 hours per year amounted to eight dollars per hour. Between 1931 and 1936, according to one survey, about 5,000 new planes were produced and 15,000 owners were registered. Approximately one-third of the owners sold their planes within a year, and 83 percent gave up their aircraft within two and a half years. In the meantime, designers and manufacturers struggled to survive.

There remained a modest but ongoing demand for a basic two-seater for instruction that offered adequate comfort and performance. Additionally, the light plane industry needed a spectrum of products to satisfy the various requirements of utility, personal business, and corporate flying. Such planes finally evolved, setting the pattern of single-engine general aviation types that lasted into the postwar era: light singles, medium-performance singles, and high-performance singles. At the close of the decade, twin-engine corporate aircraft were also introduced. A bewildering array of manufacturers introduced aircraft in all of these categories, some more successfully than others.

William T. Piper was an oilman in Bradford, Pennsylvania. In 1928, he acquired a few hundred dollars' worth of stock in Taylor Brothers Aircraft, whose guiding spirit, C.G. Taylor, had designed a couple of small sport planes with tandem seating. Taylor's "Cub," Model E-2, licensed in 1931, became the progenitor of a remarkable series of light aircraft. The E-2 cost $1,325 and annual production amounted to about 16 to 20 planes for several years. After bankruptcy of the airplane company and a split with the moody Taylor, Piper abandoned the depleted oil fields around Bradford and launched the Piper Aircraft Corporation in 1936.

In the meantime, Walter C. Jamouneau, a Rutgers graduate, had joined the Taylor company as one of its first college-trained engineers. Although Rutgers did not offer a degree specializing in aeronautics, Jamouneau had taken courses in aerodynamics and aircraft structure. In 1935, William Piper gave him the opportunity to modify the Model E-2. Jamouneau made the cockpit more weatherproof, rounded off its squared wing and tail surfaces, and increased the horsepower. The plane was awarded a different alphabetical designation, chosen for its recent stylist: the J-2. This basic design became the bread-and-butter airplane of the Piper Company. In 1936, when fire wiped out Piper's manufacturing facilities in Bradford, the company relocated in an abandoned silk mill in Lock Haven and began production of a new version of the Cub called the J-3. The refined Cub was painted bright yellow with a black "speed stripe" down the side, a scheme that made it universally recognizable. Although Piper offered fancier versions like the two-place, side-by-side Coupe and the three-place Cruiser, the classic J-3 continued as the mainstay of Piper's business until the postwar era.

Piper struggled through the lean years of the depression through a combination of enforced austerity, shrewd publicity, and an aggressive phalanx of salesman-pilots. William Piper himself kept a sharp eye on production costs and insisted that the price of Piper airplanes be kept to a minimum. In 1939, a new Cub could be had for $995 at the factory. Piper aircraft invariably showed up at air shows around the country; one act featured a Cub landing atop a platform attached to a moving automobile. The company provided eight hours of free flight training with each new plane, and dozens of aviation writers who showed up at Lock

Haven were rewarded with familiarization flights and a few hours of solo time at no cost. Such policies paid off in continuing sales. In the spring of 1940, *Fortune* magazine ran an article on the company that was building more airplanes "than any other manufacturer in the world." Production rose to more than 600 planes per year, and Piper built more than 6,000 between 1938 and 1942, when wartime demands took over the industry. Military flight training requirements accounted for many of these prewar sales, although the civil market had revived to the point where Piper and its competitors were building larger and more refined airplanes.

By 1940, pressure from competitors like Aeronca, a rejuvenated Taylorcraft, and others had prompted Piper's designers to introduce additionally modified and embellished models. With a 75–hp Lycoming engine, the three-place Piper Cruiser was definitely a cut above the tandem trainer Cub and enlarged the scope of taking to the air with partners on a business trip, or flying for pleasure with a family in a plane that still listed at under $2,000. The two-place, side-by-side 65–hp Coupe represented an overt concession to increased passenger comfort and eye appeal, both inside and out. After Piper hired an industrial stylist from Detroit, the Coupe definitely displayed automotive refinements. And why not? As one aviation journalist approvingly reported in *Air Facts*, one of the era's trade journals, "There has always been too much of a letdown in connection with . . . driving out to the airport in a well appointed, comfortable car and getting into a ship of twice the price of the car and less than half its appearance of quality." The Coupe's fully enclosed cowling sported decorative stainless steel striping across the air intake openings and over the crest of aerodynamically shaped wheel pants. Inside the cockpit, blue and gray Naugahyde upholstery on the seats and side panels set the color scheme for the chrome-accented instrument panel. And the Coupe had a muffler. "With the ice broken," the journalist declared, "this is to be heralded as the beginning of the end of airplanes so noisy they are exhausting to fly."[2]

While Piper continued to dominate the category of light singles, with about 45 percent of the market, other companies began to tap the growing demand for larger airplanes suitable for utility flying (photographic surveys, light cargo, agricultural applications) and for business flying. Before World War I, Clyde Cessna built several monoplanes and flew them in various air shows around the Midwest. During the 1920s, he formed a partnership with Walter Beech and Lloyd Stearman, and their Travel Air Manufacturing Company in Wichita, Kansas, produced several successful high-wing, enclosed-cabin aircraft. By the mid-1930s, Cessna operated on his own, joined by a nephew, Dwane Wallace, a recent graduate of Wichita University with a degree in aeronautical engineering.

The Cessna Company, supported by local investors, produced the C-34 Airmaster, a four-place monoplane. Introduced in 1934, it was one of the first American light planes to employ flaps, and after winning a series of contests in 1936, the Cessna Company advertised it as "the world's most efficient plane." Although Clyde Cessna retired in 1936, the company continued under Wallace's guidance. By 1941, when World War II halted production, a total of 186 Airmasters had been delivered, enough to keep the firm solvent.

The significance of the Airmaster and planes like it lay in the opportunity for economical business flying. Similar aircraft produced by Stinson, Howard, and Waco had to use larger and more expensive engines to get equivalent performance. With 145–165 hp power plants, the Cessnas cruised at more than 150 MPH and carried their four passengers in relative comfort on flights of 500 to 700 miles. Piper's two-place Cubs provided entry into personal flying, but planes like the Airmaster made business flying feasible. The price also reflected the clientele: the Cessna Airmaster of 1936 sold for $4,995, and the 1939 model retailed for $6,400.[3]

After a stint as a flight instructor during World War I, Walter Beech had barnstormed around the country before settling in Wichita, where he eventually became a partner with Clyde Cessna in the Travel Air Manufacturing Company in 1924. Following Cessna's departure in 1927, Beech ran Travel Air as president (his secretary, Olive Ann Mellor, later became Mrs. Beech), and the company continued to produce a series of medium-sized transports and racing planes. With 1,000 employees in 1929, Travel Air produced up to 25 planes per week. There were 95 commercial manufacturers that year who produced a total of 5,357 units, and Travel Air's contribution was 547 planes, a figure that demonstrated the firm's success in American aviation. Following the collapse of the stock market, aircraft sales nose-dived. In 1930, all manufacturers combined delivered only 1,582 aircraft. Heeding these warning signals, Beech sold out a year later to the Curtiss-Wright organization and netted a tidy profit.

Restless in his momentary retirement, Beech soon decided to reenter the airplane business. With retractable gear and a choice of engines from 225 hp to more than 400 hp, Beech's Model 17 biplane competed on highly favorable terms with other contemporary designs in its class, like the Waco and Howard aircraft, and won several competitions. The company sold Model 17 planes (known as the "Staggerwing") to pilots around the world, shipping the aircraft by sea, or, in the case of one plane, in the hold of a giant dirigible en route from Lakehurst, New Jersey, to Germany. Beginning in 1936, 20 examples of the Model 17 were license-built in Japan by the Japan Air Transport Company of Tokyo.

The machines produced by the light plane manufacturers offered prospective customers a wide range of personal aircraft, from light, two-place training and pleasure aircraft through medium- and high-performance planes. Even in the Depression, aircraft proved useful and advantageous for a wide range of business and utility flying. Variants of the Piper, Cessna, Beech, and other designs survived into the postwar period, indicative of the new standards of quality, comfort, and performance achieved during the thirties. Products like these were the archetypes of the increasingly significant general aviation industry. Still, production remained minuscule by later standards, or even in comparison with the halcyon twenties. Beech produced only 36 planes in 1935, and Cessna turned out 50 Airmasters in 1937—approximately the break-even point for the company.

Despite the slow pace of production, a slightly quickening tempo of demand in the late thirties stimulated the general aviation community to investigate other possible markets. In 1935, Beech began work on an advanced twin-engine design to carry two pilots and six passengers, which the company hoped to sell as a deluxe executive aircraft or as a small transport for feeder airlines. Two years later, the Model 18 rolled out of the hangar doors. With its dual rudders, the all-metal, 190-MPH Twin Beech bore a strong resemblance to the larger Lockheed Electra, and Beech built the new plane with techniques characteristic of companies producing larger, more complex airliners. The standard price came to $33,000. Beech's hopes of selling the plane as a feeder transport in the United States floundered, since most airlines were intent on developing major trunk routes, which required larger aircraft. Outside the United States, the Twin Beech operated efficiently on routes of lower passenger density. Export sales comprised an important share of Beech production; by the end of 1939, owners flew Beech aircraft in 23 countries around the world, and annual sales amounted to $1.3 million.[4]

Cessna also sensed the market for a light twin. For owners seeking a plane at less cost and performance just under the larger Twin Beech, Cessna designed the T-50 to sell for under $30,000 and carry five passengers at a cruising speed of about 170 MPH. Although the T-50 became available in 1939, only 43 civil aircraft had been delivered by 1942. The outbreak of war in Europe caused the U.S. Air Corps and the Royal Canadian Air Force to preempt civilian deliveries and order modified T-50 planes (respectively, the AT-17 and the Crane) as advanced twin-engine trainers and as bomber-crew trainers; the Twin Beech experienced a similar conversion for national defense. Continuing military requirements committed all of Cessna's energies, in addition to those of Beech, Piper, and others, to production for the armed services.[5]

Like their counterparts in the airline transport industry, general aviation manufacturers became a vital part of the national defense effort at the decade's end. The sophistication of general aviation designs sparked promising civil sales levels, even for the depression era. With their comfort and versatility, the improved designs of late-1930s aircraft established a quantity of orders that led to volume production and a true light plane industry. Accelerated by wartime demands, general aviation manufacturing reached industrial maturity.

Evolution of Modern Airliners

During the late 1930s, American airline transport types set the standards for international design. In an era generally characterized as one of design revolution, several features combined to give aircraft a new look and new operational success. Retractable landing gear of reliable design and operation was one such feature, and it became universally adopted following refinement of a comparatively simple device known as the O-ring. Early retractable landing gear mechanisms had to be cranked up and down by the pilot. During the early 1930s, designers tried electric motors and moved on to hydraulic cylinders in order to keep weight under control as planes and landing gear grew in size. Regrettably, hydraulic cylinders of the era used leather packing as seals, creating fluid leaks, balky operation of the gear, and expensive maintenance. In 1933, Niels Christensen devised a hard-rubber O-ring to replace troublesome leather parts. Patented in 1937, his O-ring and its housing directly contributed to the success and reliability of retractable landing gears that appeared on all sorts of planes by 1940. Many other factors contributed to the design revolution: improved engines, variable-pitch propellers, aerodynamic advances, and expansion of the aviation industry's infrastructure. At decade's end, twin-engine and large, four-engine airliners from American factories clearly led the world in performance. A major reason had to do with operational advantages in terms of payload.

A hugely significant result of the design revolution concerned a performance factor known as maximum takeoff weight, or MTOW, expressed as a percentage of an airplane's useful load (payload, fuel, operational equipment, and crew) related to its overall weight. A number of American designs in the 1930s, such as the Lockheed Orion types, boasted a useful load as high as 37 percent MTOW. A similar German design of the same era, the Heinkel 70, possessed marginally better speed, but had a 26 percent MTOW factor—a considerable shortcoming for day-to-day commercial schedules over several years of operations. In

1934, American twin-engined transports such as the Lockheed 10, Boeing 247, and Douglas DC-2 possessed useful loads of 36 percent, 34 percent, and 34 percent MTOW, respectively. As larger and more profitable planes than single-engine types, they also boasted an MTOW factor of greater value than European competitors. The significance of the MTOW factor stood out even more in the case of large, four-engine flying boats, which were the forerunners of long-range intercontinental airliners of the late 1930s. American designs from Sikorsky and Martin boasted an MTOW factor nearly twice that of European designs, but manufacturers overseas refused to learn from their American competitors. When they read U.S. manufacturers' aircraft performance figures, they dismissed them as Yankee exaggeration. One technical consultant for Britain's Imperial Airways told his client that American predictions for operating the Martin M-130 on transatlantic schedules "could not possibly be achieved."

One explanation for the disparity in performance between American and European transports had to do with geography. America's heavily travelled internal routes between New York and Chicago and on to San Francisco had no European counterparts in terms of such long distances. Moreover, competition among U.S. airlines on these and other well-traveled routes remained keen. Heavily subsidized by their governments, the European international airlines and monopolized colonial operations had less incentive to order planes for speed or profitable MTOW factors. The American advantages became translated into remarkably successful designs such as the DC-3 and the big, four-engined land planes that Douglas, Boeing, and Lockheed built for airline operations of the late 1930s and early 1940s.[6]

The evolution of the DC-3 began in a highly competitive environment, involved rapid transition based on the promise of its immediate DC-1 and DC-2 predecessors, and benefitted from international publicity generated by an intercontinental air race. In 1931, a Fokker airliner of TWA crashed into a field near Bazaar, Kansas. Among the passengers was legendary Notre Dame football coach Knute Rockne. His death shocked the nation and focused attention on the issue of safety for wood and fabric construction of planes like the Fokker Tri-Motor. TWA knew it had to find a new type of airliner, and it needed to have performance comparable to the Boeing 247 planes recently ordered by United Air Lines. TWA also knew that Boeing, part of the holding company that owned United as well, would not be able to supply them with Model 247 planes any time soon. TWA sent out a succinct, two-page letter to a list of manufacturers, asking for a new trimotor of metal construction and high performance that was obviously intended to surpass United's Boeings. From Santa Monica, California, Douglas was the only manufacturer to reply. The company had subsisted on military contracts, and reduced orders during the Depression underscored the need to diversify into commercial markets such as airline transports. As the Douglas design team pondered its response to TWA's prospectus, they decided to propose a twin-engine design to avert resemblance to the discredited trimotor style and to compete with Boeing. When Douglas executives headed for the East Coast to make a formal presentation, the design remained incomplete; they worked on final details en route. This was easy enough to do and they had plenty of time because they traveled by train—an ironic commentary on the state of coast-to-coast business air travel in the early 1930s.

The Douglas team successfully sold TWA on their proposal for the DC-1 (for Douglas Commercial No. 1), a plane that featured more power than the Boeing 247, carried 12 passengers, and used aerodynamic features such as improved NACA cowlings and wing construction, the details of which had been refined in Cal Tech's new wind tunnel. In the fall of 1932, Douglas got a contract for $125,000 to build a prototype; TWA took options for up to 60 planes at $58,000 each. During February 1933, the Boeing 247 made its first flight while the first DC-1 was under construction at the Douglas plant.

The DC-1 took off for the first time in July 1, 1933, and its flight trials left TWA highly optimistic about its possibilities. Already, Douglas realized that an addition of a foot and a half to the fuselage plus some internal modifications would permit 14 seats instead of 12—almost 50 percent more than the 10 seats in Boeing's 247. More powerful engines would boost the cruising speed to 196 MPH. TWA enthusiastically agreed to the changes. Consequently, the prototype DC-1 remained the only one of its kind, with the DC-2 configuration going into production as TWA's new airliner.

The DC-2 entered service during the spring of 1934, setting new records for speed between cities, then breaking its own records. Transcontinental service between Newark and Los Angeles took 18 hours with scheduled stops at Chicago, Kansas City, and Albuquerque—a schedule that eclipsed the 247's record by more than one and a half hours. The DC-2 flew faster, carried more passengers, and its larger fuselage and uncluttered main aisle made a convincing contrast to the 247's cramped cabin with a wing spar to climb over. Within a year, every major airline in the United States had the DC-2 on order except for United, whose corporate parentage dictated retention of its Boeing legacy.[7] Douglas had clearly conquered the domestic market. Success in the international arena soon followed, giving the Douglas design global recognition as the harbinger for a new epoch of air travel. In the process, American airline transport technology continued to evolve as the world's standard.

Much of the renown of the DC-2/3 series can be attributed to Australian philanthropy in the form of the MacRobertson Confectionary Manufacturers, and to Dutch interest in developing its airline route to the Dutch East Indies. In America, the MacRobertson England to Australia Air Race of 1934 is not always remembered as well as the Schneider Cup competition or the Thompson races of the 1930s. But the MacRobertson derby became one of the most compelling events in the history of the American aviation industry. Although the prize money was tempting, it was the international character of the race and its awesome geographic challenge that drew an impressive roster of 20 entrants from Australia, Britain, Denmark, Holland, New Zealand, and the United States. One of the first entries, sponsored by the Warner Brothers film producers of Hollywood, was a Boeing 247 piloted by the dashing Roscoe Turner. The international Dutch airline, KLM, announced a Dutch crew would fly one of the line's new Douglas DC-2 transports, the production version of the prototype DC-1. The British had no airliner capable of matching the fast American planes. In fact, the British found it necessary to design a new plane from scratch in order to remain at all competitive. This was the de Havilland D.H. 88 Comet, a streamlined, twin-engine monoplane with just enough room for a pilot and navigator seated in tandem. Three were built to carry Britain's colors.

The de Havilland Comet's crew was geared for utmost speed, enduring a cramped cockpit and wearing the racer's uniform of helmets, goggles, and coveralls. They were also pushing their plane's engines to the limit in order to stay ahead of the Dutch in the DC-2. During the stop in India, the British pilots were obviously tiring and nervous about the punishing pace on their finely tuned aircraft. In contrast, a race official clearly remembered the almost casual Indian stop made by the Dutch commercial transport and its crew, only four hours behind. "It was most impressive in its simplicity. A circuit, a landing, a reporting, a meal, a refueling, a takeoff. There were no histrionics. The pilots wore their usual KLM uniforms and the quality of their airmanship made a deep impression on me." Moreover, press reports noted that the Dutch airliner carried along three paying passengers in addition to its two-man crew. By the time that the de Havilland Comet landed in Australia as the winner, the comparatively huge DC-2 was nipping at the heels of the compact British racer. Awarded the handicap prize, the DC-2 had nearly taken it all. A writer for the British *Saturday Review* underscored the significance of the Douglas airliner's performance. Britain's special, two-place custom racer might have won, but "close behind . . . thundered an American Douglas machine with a seating capacity of 20 persons," the correspondent observed. The article emphasized that no British airliner—not even any plane in the Royal Air Force—could have finished within 1,000 miles of the American transport. "It is almost incredible," the writer admitted, "but it is true."[8]

KLM not only developed its international routes but also became a European leader. More importantly, the MacRobertson race triggered a spurt of orders for Douglas transports, including a dozen more from KLM, where DC-2 liners on its routes to the Far East became synonymous with fast, efficient service. The DC-2's impressive performance won it orders from a variety of other European airlines as well as from Australia and China. Perhaps its closest competitor was the Lockheed Electra (10 passengers), also flown by KLM in the West Indies and by operators in Europe, Canada, South America, and the Pacific region. Aside from the plane's inherently valuable qualities, the international notoriety generated by the MacRobertson race played no small part in its success.

While the DC-2 created a sensation in the European press, its even more historically significant successor began to take shape. Having seized an obvious advantage over rival Boeing in the development of commercial airliners, Douglas proceeded to consolidate its lead by producing the classic DC-3. Outwardly similar to its predecessor, the DC-3 had a larger cabin and wingspan. A pair of 1,000-hp engines meant increased speed and payloads. With its "day-coach" capacity for 21 passengers, coupled with overall aerodynamic efficiency, the DC-3 boasted seat-mile costs as much as one-third to one-half lower than its contemporaries. In the spring of 1936, American Airlines put its DC-3s into regular service on the New York-Chicago run, inaugurating the first airline operations that made money just by hauling passengers and freeing companies from total dependence on government mail contracts to make a profit. Two years later, 80 percent of the U.S. airline fleet were DC-3s, and more than two dozen foreign airlines were operating the Douglas transport as well; Japan and Russia built them under license. Again, KLM was among the first to place orders from overseas, as domestic and foreign orders for the DC-3 began to multiply. Eventually, some 11,000 civil and military versions were built, with venerable DC-3 transports still flying commercially into the 1990s. The DC-3 design became one of the icons of twentieth-century progress. It not only embodied a synthesis of the best of the prewar era's engineering but also represented the aerodynamic, streamlined motif that influenced Art Deco styling and industrial design of later years.

Sales did not translate into immediate profits, however. During the late 1920s and early 1930s, Douglas had prospered by concentrating on military products. Between 1925 and 1931, the company sold 814 aircraft, only 14 of them to civil customers. The Post Office Department bought 51 mail planes. A total of 635 went to the Army and 78 more to the Navy; 34 more

The majority of Douglas DC-3 transports in military service carried the designation C-47. Having entered airline service in 1936, the civilian version became the standard American commercial airliner and equipped many foreign companies as well. Wartime industrial experience helped American companies win a larger share of the postwar civil market. *Courtesy of the McDonnell Douglas Corporation.*

went to China, Mexico, and Peru, largely as military exports. By 1934, success of the DC-2 and DC-3 gave preeminence to commercial transports, and military orders sank to about 15 percent of production. Even so, Douglas lost money on commercial sales up to the start of the war.[9]

Rising passenger statistics in the U.S. encouraged airlines to order four-engined transports to carry twice as many passengers as the DC-3. This trend, to be sure, had been anticipated by the huge four-engined flying boats of the late 1930s,[10] although their transoceanic routes and standards of service imparted a different aura of worldly chic. German dirigibles had pioneered transoceanic passenger flights in the mid-1930s, a fascinating venture that dramatically closed with the *Hindenburg* disaster in 1937. Flying boats succeeded them,[11] competing directly with steamship lines for a well-heeled clientele. Pan American's Martin and Boeing Clippers offered sophisticated service, vintage wines, and haute cuisine.

They incorporated a high degree of structural sophistication. The end of the Clipper era came with the development of a new generation of long-range, four-engine airliners. Their enhanced performance and reliability undercut the argument for seaplanes on over-water routes, and the proliferation of global military air routes during World War II provided a reassuring number of alternate airports within emergency range. Other significant features included pressurization.

The Army had begun research on the design of pressurized cockpits in 1935 and carried out pioneering flight tests in 1937 with the X-35, a completely pressurized version of Lockheed's Model 12 airliner (similar to the Electra and the Lodestar). The trend in commercial airliners began with the Boeing 307; pressurization permitted it to overfly threatening weather and dangerous storms. The 307 entered service in 1940, followed by the unpressurized Douglas DC-4, which first flew in 1942, and the pressurized Lockheed Constellation in 1943. Pressed into military service because of the war, the new airliners did not make their full impact felt in civilian service until 1945. With more powerful engines, additional passenger capacity, higher speeds, longer range, and the ability to fly above bad weather, variations of these four-engined airliners made commercial transoceanic flying comfortable, affordable, and commonplace in the postwar era. In his informed study on *The Modern Airliner*, Peter Brooks categorized such hallmarks as being representative of "the DC-4 generation."[12]

The original DC-4E, funded by five airlines—American, Eastern, Pan American, TWA, and United—rolled out of the Douglas plant in 1938 and made demonstration flights over the United system in 1939 but never entered production. Rising costs and the chance to get pressurized Boeing 307 Stratoliners broke up the five-company consortium. The development costs totaled some $3 million. Some of the escalating expenses were due to the complexities of developing much larger, four-engined transports. Others arose from the introduction of new production techniques, such as flush riveting, designed to get thousands of rivet heads out of the airstream in order to reduce drag. Flush riveting had been around since the early 1930s; Boeing employed the technique on its P-26 fighter. But many applications were limited, and the Douglas effort to use the technique over the entire plane seems to have stimulated general use of flush riveting throughout the industry. With Edward P. Warner as consultant, Douglas also pioneered the design of planes with built-in "flying qualities" for better handling in the air.

The DC-4E reflected a number of other innovations that characterized the new transports and demonstrated increased attention to passenger appeal as the manufacturers courted airline customers. Marketing, in other words, played a stronger role. Douglas press releases at the time stressed the tricycle landing gear, a first for planes of this size, and dubbed it the "Tri-Safety landing gear." The tricycle undercarriage had appeared on other planes before World War II, but its use on the new generation of heavier airliners contributed to their efficient operation during takeoff and landing. Moreover, luggage-laden passengers who had scrambled up the aisles of "tail draggers" like the DC-3 enjoyed arriving aboard the newer planes with aisles and seats horizontal. Douglas claimed that the DC-4E fuselage was larger than a rail coach—one of the plane's many features for passenger comfort that became standard in the future. Company brochures listed complete soundproofing, steam heat, hot and cold running water, air conditioning, and lavatories large enough for changing clothes. In the trimotor era, professional interior consultants with qualifications in industrial design seem to have been enlisted as an afterthought. By the time of the DC-4E, the well-known industrial design firm of Howard Ketcham, Incorporated was brought in early in the game as a standard operating procedure. The success of the "DC-4 generation" owed much to these collective refinements.[13]

The Boeing Stratoliner, Model 307, used the wings, tail assembly, and some systems from the B-17C Flying Fortress. Development began in 1935, and available components led to first flights in 1938, followed by deliveries to Pan Am in 1940. The first four-engined pressurized airliner to be built in America, the Stratoliner allowed airline navigators to set a course at altitudes where the avoidance of rough weather meant less discomfort for passengers and better adherence to airline schedules.[14] An accident threw a pall over the Stratoliner's sales potential and World War II put an end to its development.

In 1939, Lockheed began work on a 40-passenger airliner, the L-049 Constellation, based on an order from TWA. The elegant, triple-tailed Constellation epitomized the era's most advanced trends: pressurized cabin, tricycle landing gear, ultra-modern cabin appointments, and so on. The early prototypes featured designer-type interiors that boldly displayed a world map against the rear cabin bulkhead to signify the plane's suitability for global routes. When America entered World War II, the Air Force took over the first production batch for service as C-69 transports. The first plane did not fly until January 1943, and the war ended before any of the early production models began regular service on overseas routes.[15]

The DC-3, DC-4, and their counterparts of the 1930s from Lockheed and Boeing served with distinction as military transports in World War II. It was a tribute to technology and training that planes and personnel shifted so effectively from civil operations to wartime demands and back again after 1945.

Research and Development for War: The NACA

Research and development involving aerodynamic theory, airframes, power plants, and other features evolved from private projects and corporate efforts as well as investigations sponsored by federal agencies. American industry also enjoyed significant benefits from NACA research during the 1930s and during World War II. Civil aircraft had used various NACA refinements to yield the performance and economy that made the DC-3 a commercial success in the United States as well as overseas. The eventual success of American combat planes in World War II gave the aviation industry an additional luster that translated into public support for Cold War budgets and profitable postwar exports.

As in the past, this success story was a mix of indigenous and foreign themes. Theodore von Karman's influence on the West Coast extended from academic pursuits at Cal Tech to his role as high-level military advisor. Through von Karman and others, the continuous flow of European information to the United States influenced many engineering faculty as well as federal researchers like Eastman Jacobs, a senior engineer at the NACA labs in Virginia.[16]

Following a trip to Europe in 1935 to attend an aeronautical conference, Jacobs acquired insights that played a central role in guiding NACA research in the theory of laminar flow applied to wing designs for high speed flight. The technique was applied to the North American P-51 Mustang, as well as other aircraft. A variety of key NACA reports began to emerge from an impressive number of tunnels that went into operation during the late 1920s and 1930s. Langley's research in its refrigerated tunnel contributed to successful deicing equipment that not only enabled airliners to keep better schedules in the 1930s but also enabled World War II combat planes to survive many encounters with bad weather. The design revolution leading to all-metal monoplane transports had a similar impact on military aircraft. During 1935, Boeing began flight tests of its huge, four-engine Model 299, the prototype for the B-17 Flying Fortress of the Second World War. The big airplane's performance exceeded expectations, due in no small part to design features pioneered by the NACA. The Boeing Company sent a letter of appreciation to the NACA for specific contributions to design of the plane's flaps, airfoil, and engine cowlings. The letter concluded, "[I]t appears your organization can claim a considerable share in the success of this particular design. And we hope that you will continue to send us your 'hot dope' from time to time. We lean rather heavily on the Committee for help in improving our work."[17]

Although the Fascist powers continued to develop civil aircraft during the 1930s, it became apparent that military research absorbed the lion's share of their attention. In response, the NACA formed stronger alliances with the U.S. military. In 1936, the agency put together a special committee on the relationship of NACA to national defense in time of war. Following congressional debate, the NACA received money for expanded facilities at Langley along with a new laboratory at Moffett Field, south of San Francisco. The official authorization came in August 1939; only a few weeks later German planes, tanks, and troops invaded Poland. World War II had begun. The outbreak of war in Europe, coupled with additional warnings from the NACA committees and from experts about American preparedness, triggered support for a third research center. Congress quickly responded, and an "Aircraft Engine Research Laboratory" was set up near the municipal airport in Cleveland, Ohio. This third new facility in the midwest gave the NACA a geographical balance, and the location also put it in a region that already had significant ties to the power plant industry.[18]

Over the course of the war years, the NACA's relationship with industry went through a fundamental change. Since its inception, the agency refused to have an industry representative sit on the main committee, fearing that industry influence would make the NACA into a "consulting service." But the need to respond to industry goals in the emergency atmosphere of war led to a change in policy. The shift came in 1939, when George Mead became vice-chairman of the NACA and president of the United Aircraft Corporation. Mead's position in the NACA, considering his high-level corporate connections, heralded a new trend. During the war, dozens of corporate representatives descended on Langley to observe and actually assist in testing. In the process, they forged additional direct links between the NACA and aeronautical industries.[19]

Military Aviation

The overall picture of the aviation business in the depression era looked surprisingly good. Between 1934 and 1938, profits as a percentage of sales started at zero and rose to 11 percent; return on investment rose from 3 percent in 1935 to 15 percent in 1938. Still, comptrollers in the various corporations remained nervous. The larger and more complex transports and bombers incurred high development costs, and nearly all of these expenses had to be absorbed by the manufacturers. Military orders continued during the 1930s, although their pace was slow. The aeronautical companies worked with fixed-price contracts, and unexpected technical problems or a shortened production run could easily mean a financial disaster. More often than not, exports of airplanes and parts provided the income that accounted for the difference between profit and loss. In the decade after Lindbergh's flight, annual exports of planes jumped from 63 to 631, and the total number of all aeronautical products (planes and parts) rose from $9.1 million in 1929 to $39.4 million in 1937. For the latter year, exports produced 26 percent of the industry's total sales.

Sales to foreign governments were negotiated without the profit limitations imposed by the U.S. government. In fact, the Bureau of Foreign and Domestic Commerce, an agency of the Commerce Department, began to push foreign aircraft sales during the late 1920s. Some American officials worried about the loss of American technology, but the military planes sold during this period were generally not of first-line quality, and they invariably went to customers in Latin America and Asia that did not have the industrial base to put their own versions into production. Competition for sales became intense in the 1930s, and one aviation figure from Wright Aeronautical admitted during a congressional hearing (the Nye Committee, on the sale of armaments) that commissions paid to many foreign nationals could be construed as bribery.

Given the plight of America's airpower as a result of slender budgets and obtuse congressional muddling, the aviation industry nonetheless possessed valuable legacies from which it could build for the future, based on its own record of innovations, NACA activities, advanced commercial designs, and resourceful individuals. Despite gaps, developments of the 1930s provided enough momentum to produce huge numbers of effective combat planes during World War II.[20]

In 1936, after years of operating biplanes from the decks of its aircraft carriers, the U.S. Navy ordered its first monoplane, the Brewster XF2A-1 Buffalo. Producing a low wing design with retractable gear, the Buffalo's designers obviously thought the time had come for a naval fighter possessing the attributes of the modern, land-based fighters of the era, and Brewster's aeroengineers deserve credit for helping to push the Navy past its old traditions.

Grumman, traditional supplier of Navy fighters, had been designing its XF4F-1 as another biplane when news of the Navy's order for Brewster monoplanes arrived in the company's offices. Quickly scrapping their standard design, Grumman's XF4F-2 suddenly became a monoplane. The wing of the Wildcat, as it became known, used stressed skin construction, featured a new NACA airfoil section, and incorporated an advanced type of landing flaps. Later designs had a folding wing system, patented by Roy Grumman, to facilitate stowage of additional planes aboard smaller carriers. Production versions of the Wildcat fought through the end of World War II, with some 2,000 produced by Grumman and nearly 6,000 more built by General Motors.

Grumman's next fighter was the F6F Hellcat. Entering production in 1943, a total of 12,275 Hellcats reached service in only 30 months, constituting a record number for so short a time, and indicative of the adroit manufacturing ability of the aviation industry. In its race to erect buildings for Hellcat production on Long Island, Grumman bought thousands of steel girders from New York's elevated railroad, plus tons of additional steel beams from the dismantled structures used for the 1939 New York World's Fair.[21]

The U.S. Navy operated numerous types of fighters, torpedo/dive bombers, seaplanes, and other aircraft during the war, but the U.S. Air Force claimed the lion's share of wartime production. With retractable gear, twin-engine bombers like the Boeing B-9 marked important steps ahead for the U.S. Air Corps. Nonetheless, Boeing's design retained an open cockpit and crew positions, and even though its cruise speed of 165 MPH was good for 1932, the winning design was the Martin B-10, whose top speed of more than 200 MPH exceeded that of the Army's pursuit planes. The initial B-10 production versions were ordered in January 1933, a total of 48 units at $54,840 each. Martin eventually received orders for more than 200 planes. All crew positions were enclosed; the front gunner sat in a transparent, rotating turret; the bomb load was internally stowed; the bombardier used early versions of the top-secret Norden bomb sight.

The B-10's performance prompted the Army to push development of faster monoplanes to replace biplane fighters in service. Trials of monoplane pursuits having retractable gear occurred during the early 1930s, although the first orders for a monoplane design went to the Boeing P-26, with fixed gear and open cockpit. Boeing's contract called for 111 pursuits at a price of $9,999 apiece, not including government furnished equipment like instruments and engine. The P-26, with a top speed of about 230 MPH, entered service in 1934. Compared to a bomber like the B-10, with enclosed cockpit and retractable gear, the P-26 pursuit plane clearly represented an interim solution.

The first all-metal U.S. Army fighter with enclosed cockpit and retractable gear was the Seversky P-35, developed by Russian émigrés Alexander P. de Seversky and his chief engineer, Alexander Kartveli. The P-35 entered service in 1938, and some versions reached 290 MPH. Born in Tiflis, Russia, in 1896, Kartveli received his aeronautical training in Europe before joining the Seversky Corporation in 1931. Corporate disagreements led to de Seversky's departure in 1939, although Kartveli stayed on. As chief designer for the reorganized firm, Republic Aviation

Corporation, Kartveli led the design team responsible for the outstanding P-47 Thunderbolt of World War II. In the meantime, other U.S. military planes were progressing through the design phase and into flight tests. Several different twin-engine attack planes evolved, along with modern four-engine bombers like the Boeing B-17, Convair B-24, and Boeing B-29.[22]

The power, complexity, and expense of the big new bombers made them controversial from the start. In the late 1930s, aerial bombing attacks in Ethiopia, Spain, and China created shocking images of urban destruction which, in the eyes of many critics, discouraged development of large bombers. At the same time, the existence of such formidable weapons and their dramatic raids on urban areas led to arguments that America also needed such equipment as a matter of deterrence in a warlike world.

Based on the promise of the B-9 and B-10 performance figures, the development branch of the Air Corps, known as the Materiel Division at Wright Field, Ohio, pushed for a bold new design to test the limits of large bomber design. Several Air Corps generals doggedly pushed the project, orchestrating some budgetary legerdemain to continue development of what became the B-17 after a prototype, the Model 299, crashed in 1935. Over the next three years, the fate of the B-17 continued to hang in the balance, waxing and waning through a series of budget fights, debates over strategy, and outright hostility from a variety of sources within the U.S. Army.

In the end, the rising military crisis in Europe resolved the issue, especially after the Munich conference of 1938, when Britain and France acceded to Adolf Hitler's territorial demands. President Roosevelt weighed reports from State Department and U.S. defense officials and heard specific warnings about the role of the German *Luftwaffe*. In November, Roosevelt ordered a preparedness program and specified an air force of some 10,000 planes. Major General Arnold, who assumed command of the U.S. Army Air Corps in 1939, referred to FDR's decision as the Magna Carta for the Air Corps. As the Air Corps plotted production for 10,000 planes, the B-17 figured heavily in its finances. For 552 copies of the bomber, the Air Corps committed more than 27 percent of its total budget.[23]

As the U.S. struggled to come to grips with the escalating world crisis of the late 1930s, the official Air Force history of World War II observed that the aircraft industry "received as much stimulus from external as well as internal sources . . . ," especially from the European democracies. When British and French purchasing commissions, scrambling to contend with Nazi expansionism, arrived on American shores in

1938, the results were electrifying.[24] During 1938 and 1939, European orders accounted for $400 million of a backlog totaling $680 million. During the spring of 1940, the Anglo-French Purchasing Commission ordered 6,000 planes from eight major suppliers. (In contrast, Congress in 1939 had approved an air force totaling only 5,500 planes, the result of legislative limits for authorized Air Corps strength as well as budgetary constraints.) After the fall of France, Britain picked up the suddenly orphaned French contracts and raised their own sights as well. By September, the RAF had issued American contracts for 14,000 planes and 25,000 engines totaling $1.5 billion. By the end of 1940, American firms had spent $83 million of their own funds on new plants and equipment. Significantly, an additional $74 million for manufacturing plants came from Britain alone. After the war, official Air Force assessments concluded that the infusion of British and French orders, which fueled a broad expansion of plants and tooling, had the effect of advancing aircraft production in the United States by a full 12 months.[25]

In one celebrated instance, European orders resulted in one of the most effective American fighters of World War II, the North American P-51 Mustang. The plane's genesis demonstrated the important role played by individuals in arguing for improved American combat aircraft, the significance of the European catalyst in America's production build-up, the legacy of the NACA, and the continuing subtheme of European-born technicians as important figures in the American industry. The story of the P-51 also exemplifies the eclectic process by which designs often emerged.

Despite the general doldrums in fighter aircraft development for the Air Corps during the 1930s, Colonel Oliver Echols kept useful work alive at Wright Field in Ohio, where he presided over an experimental section that funded a small but consistent stream of new designs for flight trials. One effort involved close cooperation with Curtiss-Wright, where designer Don Berlin directed work on an advanced fighter, the XP-46, with an in-line, liquid cooled engine. The plane's distinctive design featured a radiator in the ventral position aft of the wings in a special configuration that contributed to reduced aerodynamic drag.

After the German invasion of Poland in September 1939, purchasing teams from Britain and France arrived in America with a heightened sense of urgency. Production of the P-40 assumed high priority for the Air Corps as well as the Europeans; the latter began scouting for other manufacturers to put the Curtiss fighter into production for them. The British Purchasing Commission proposed that North American undertake P-40 production, but the company disliked the idea. At this point, Colonel Echols suggested that the commission fund a contract with

North American to develop a totally different combat plane, using data from the Air Corps and Curtiss-Wright that related to the XP-46 project. During May 1940, North American negotiated the sale of XP-46 data from Curtiss Wright for $56,000 and also finalized a Foreign Release Agreement with the Air Corps. This cleared the way for North American to receive relevant test reports from Wright Field; in return, North American agreed to deliver two production models of the new fighter to the Air Corps for evaluation. With bureaucratic sleight of hand and European funds, Colonel Echols procured another potential combat plane for the American armed forces.

On May 23, 1940, North American and the Anglo-French Purchasing Commission signed the formal contract for 400 examples of a plane called the Model NA-73. Working seven days a week, North American's design team melded Curtiss-Wright's concepts with their own ideas, and decided to use a laminar flow airfoil section recently reported on by the NACA. When the NACA's Eastman Jacobs stopped by North American during a visit to California, the design team hustled him off to their drawing boards for advice on blending the new airfoil into their plane.

The plane's design, obviously a collective effort, reflected considerable influence from two individuals in particular. J. Leland (Lee) Atwood, Chief Engineer in 1940, pushed the idea of using the aft-mounted radiator from the XP-46 program. The assistant chief engineer, Ed Schmued, acted as principal within the preliminary design group and, as one knowledgeable engineer noted, "in all probability Schmued would have done the lion's share of the work." As such, Schmued embodied the persistent line of European influence in American aviation development. Born and educated in Germany, he emigrated to Brazil, where he was involved in various aviation ventures and became a service representative for General Motors. Arriving in America in 1930, he became a design engineer for the General Aviation Manufacturing Corporation, a Fokker licensee, and eventually wound up at North American during the corporate shuffling of General Motors and other holding companies. Writing to a Rolls-Royce historian some four decades later, Schumed remarked that "[t]he P-51 was the first airplane I designed."[26]

Keeping in mind the need for rapid development and production, the NA-73 design group decided to use existing internal components if they could. As a result, the NA-73 incorporated several features from North American's AT-6 trainer, including electrical system, hydraulics, wheels, and brakes. To facilitate mass production, the design team specified squared-cut wingtips and tail surfaces in place of the rounded shapes characteristic of the era. The NA-73 rolled out of North American's design shop in only 102 days; the Allison V-1710 engine arrived about three weeks later; the Mustang became airborne on October 26, 1940. Without detracting from the role of North American's design team, the Mustang's balanced design and rapid evolution certainly owed much to the early work of Curtiss-Wright and the enterprise of Colonel Oliver Echols.

In the course of 1941, the P-51 Mustang received additional orders from the Air Corps, and the first units arrived in Britain, where the plane reached duty squadrons early in 1942, followed by combat sorties that May. Pilots liked the Mustang, although the Allison engine left it underpowered at high altitudes, so that the RAF and USAF both assigned the P-51 to ground attack and tactical reconnaissance roles. Meanwhile, a shrewd test pilot from Rolls-Royce, Ronald W. Harker, made a trial flight in the Mustang and immediately realized that a recently improved engine would make it more than a match for the latest German fighter threat, the Focke Wulf FW 190. Reequipped with a new version of the Rolls-Royce Merlin engine, the Mustang responded like a thoroughbred, performing strongly at high altitudes above 20,000 feet. Working closely with the Air Force, North American, and Packard (who built the Merlin under license in America), British and American engineers resolved modification problems and the P-51B emerged as one of the finest combat planes of the war. About 3,000 eventually went to the RAF; more than 12,000 served with the U.S. Air Force* during World War II.[27]

The wartime experience transformed the old ways of doing business in more ways than one. In addition to the rapid expansion in terms of size, many manufacturers learned to do more business overseas. Lockheed, for example, employed a few hundred people in 1938 and had produced several dozen planes during its entire corporate lifetime. By 1945, the company rolled out 23 planes per day, employed 90,000 people, and held war contracts valued at $2 billion. Lockheed also became a multinational corporation. Already in 1938, its marketing efforts extended to Europe, where its twin-engine airliners operated, and where the company touted military versions of Hudson and Ventura transports. During 1938, Britain placed a $65 million order for these aircraft, with options on the P-38 fighter. Late in the war Lockheed listed operations in England, Northern Ireland, Scotland, and Australia. Multinational activities like these helped set the stage for much more complex arrangements in the postwar era.[28]

---

*The Army Air Corps became the U.S. Army Air Force in June 1941; it was generally called the Air Force or USAF.

World War II: Politics and Production

The European momentum, strong as it seemed, disappeared into the tidal wave of production eventually achieved by American manufacturers. Alarmed by the rapid progress of the German campaigns during the spring of 1940, Roosevelt's advisors put a high priority on the rapid development of American airpower. In the spring of 1940, the President announced the startling goal of building 50,000 planes per year. The government created a National Defense Advisory Commission, with William S. Knudsen, president of General Motors, on a leave of absence from his GM position to take charge of production. A yawning chasm existed between Roosevelt's goals and production reality. Existing capacity represented about one-fifth of what was needed for airframes, engines, and propellers, and the total output of planes for 1940 came to 12,813, about one-half of which were military designs. Nonetheless, within two years American industry reached Roosevelt's goal, with bigger agendas already projected. Success was achieved through the efforts of individual companies to raise stock; multimillion-dollar loans from the Depression-era Reconstruction Finance Corporation; innovative funding and support through federal bureaus such as the Defense Plant Corporation; cooperation between rival manufacturers to achieve maximum production for specific designs; and hard work by everyone involved, including aviation-related unions.

Through the 1930s, unionization remained unpredictable. Most of the aviation manufacturers had company unions, and corporate directors successfully beat back outside union efforts. From 1916, the International Association of Machinists (IAM) made occasional forays to enlist members, but its narrow preference for crafts proved unattractive to aviation workers who considered themselves aircraft builders rather than separate cohorts of electricians, machinists, and so on. Other national unions generally failed to win support from the rank and file of aviation workers for the same reason.

Responding to patriotic exhortations, the American Federation of Labor and the Congress of Industrial Organizations pledged not to strike for the duration of the war. Inevitably, a few rogue strikes occurred. Wage disputes and work rules that carried over from the 1930s engendered collusion by many manufacturers to discourage unions and harass activists, but productivity usually remained high because workers took pride in jobs in what they perceived as the era's most advanced industrial sector.[29]

In creating a new industry for aviation manufacturing, military leaders learned the value of good publicity and how to manage it. To a certain degree, the mass-production miracles of the war relied on successful propaganda to support industrial expansion of new and unfamiliar technologies. There was not necessarily inherent support by the public or by Congress for the huge outlays of money required to develop an instrument of airpower—a concept that was still an abstract theory for most Americans, regardless of events in Europe and Asia since 1936. This tepid response persisted during the slow buildup of American air forces in Europe during the early years of U.S. participation. Support for expensive mass production wavered when stories about the early inadequacies of American designs began to surface. Some critics demanded that American manufacturers should produce British designs, a procedure that would have enriched British manufacturers at the expense of American industry. These issues helped explain the intensity of the furor surrounding Alexander de Seversky's book, *Victory through Air Power*, published by Simon and Schuster in 1942. In the book, Seversky roundly criticized the combat worthiness of American warplanes in general, questioned the effectiveness of U.S. Navy aviation, sniped at French and British responses to German aviation, and appealed for massive production programs of advanced designs as a matter of survival in the face of modern airpower.

Victory through Air Power achieved remarkable success; with 350,000 copies in hardcover and softcover editions, it was one of the most widely sold books of the World War II era. Moreover, *Reader's Digest* produced a condensed version, and this endorsement alone assured a wide readership for many thousands of Americans who might have easily dismissed such a title. For those who missed these publications, a number of newspapers across the nation carried installments of de Seversky's tract. As if this were not enough, Walt Disney waded in with an animated version, in which de Seversky's explicit warnings became visually translated for presentation on motion picture screens. Released in 1943, the Disney treatment attracted favorable reviews, although it was not a box-office hit.[30]

Air Force leaders were seriously vexed about the apparent criticism of aerial doctrine at a time when they were attempting to muster support for the production of thousands of aircraft for training and combat. At the same time, the public's response to the book actually furthered their aims. The book's title became a popular slogan, mirroring the resolve of the American public to produce a multitude of planes capable of defeating the enemy. Ultimately, de Seversky's book and the attendant publicity made a significant contribution to the war effort.[31]

The labor force attained its peak toward the end of 1943, totalling 2,100,000 men and women. The latter group comprised some 40 percent of the workforce in the Los Angeles area, a proportion that was echoed in many centers across the country. Although women had been involved in aircraft manufacturing since the World War I period—often as seamstresses to sew fabric covering of the era—they entered aviation trades by the thousands during World War II. Many were teenagers, others were housewives; the West Coast had a large share of blacks and Latinos. Whether they had held prior jobs as clerks, hairdressers, gas station attendants, or domestics, the women generally faced varying degrees of hostility and sexual harassment. Still, they learned their tasks well, and many rose to supervisory roles on assembly-line production teams. The women often took jobs over the opposition of spouses and families; although most returned to family roles when peace ended the massive wartime employment, the evidence suggests that they were more assertive in their postwar lives and encouraged their daughters to become more independent. More black women who had left domestic work to take wartime jobs in the aircraft plants seemed less inclined to return to former employment, but sought positions in durable goods industries after the war, with better wages and more security. In these respects, the wartime aviation industry involved notable social changes for large numbers of women. The same held true for nonwhite males who entered wartime aviation trades. The struggle to retain skilled jobs at higher pay remained at issue until affirmative action legislation became more effective in the mid-1960s.[32]

The astonishing aviation production records of American industry accrued from many sources. Efficient federal planning, the contributions of private industry, and the skills of thousands of men and women merged into an admirably effective partnership. A variety of management theories and styles inevitably evolved. One exasperating problem involved a means of translating the nation's production goals into realistic requirements for floor space, man-hours of labor, and related factors necessary to produce a given weight of airframe, whether it was a four-engine bomber, advanced fighter, or basic trainer. Theodore P. Wright, director of engineering for Curtiss-Wright at the start of the war, proposed such a statistical formula in 1939. The concept, first published in *Aviation* magazine, appears to have guided subsequent Air Force planning for successful mobilization and production programs. Another challenge involved problems of subcontracting with suppliers large and small across the nation, many having little or no experience in the aircraft industry. Special committees evolved under the aegis of the Production Division at Wright Field, with the prime contractors establishing selected engineering subcommittees to iron out potential snags.[33]

In addition, the demand for increased production capacity required huge new plants in new locations. With local risk capital and federal funding, the Dallas area became the site of a major facility for North American. Built on cotton fields near the small town of Grand Prairie, population 2,000, the new plant employed 39,000 people by the end of

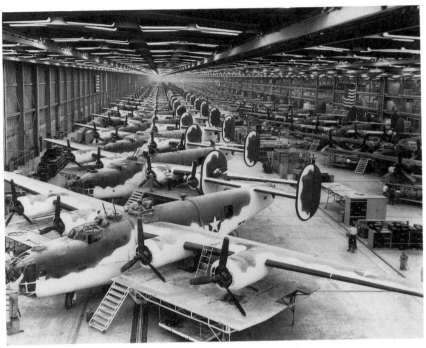

Seen in one of the brand-new plants built for mass production during World War II, personnel in Fort Worth, Texas, complete final work on Consolidated B-24 Liberator long-range bombers. The crowded production lines of American manufacturers symbolized one of the most significant contributions to the eventual Allied victory. *Courtesy of General Dynamics.*

the war. The influx of such numbers created severe housing problems, even though the Federal Works Administration built 300 new homes for a workers' subdivision known as "Avion." The main production facility, with 855,000 square feet of floor space, went up in seven months during 1940–41, and was touted as the first air-conditioned and artificially lighted aircraft plant designed to operate 24 hours per day. Between 1941 and 1945, North American workers built 20,000 planes there, mostly P-51 Mustang fighters and AT-6 Texan trainers. Another huge plant took shape in Fort Worth, where Consolidated Aircraft built a mammoth, elongated plant that turned out over 3,000 B-24 Liberator bombers in three years.[34]

The dispersal of plants for security reasons played a role in early plans, but the principal determinant became the availability of labor, along with housing, power, and transportation facilities. Inevitably, this led to new aviation production lines for airframes and engines in the

Detroit area, where planners assumed that the American genius in mass producing autos would be applied to aircraft. The principal example of such assumptions became Ford's plant complex at Willow Run. The project turned out to be costly, time-consuming, and frustrating, leading to the caustic nickname "Will it run." Beginning in May 1941, the huge assembly lines and their complement of sophisticated production machinery were designed to turn out the Consolidated B-24 Liberator bomber. Snags occurred immediately, when Ford's production engineers uncovered discrepancies in Consolidated's general plans and detailed drawings for parts—in the rush to production, Consolidated officials had counted on their own skilled assembly-line supervisors to rectify such problems.

Once Ford got its operation humming, production reached 309 units per month, plus parts for about 112 more planes. At its peak in 1944, Willow Run's standard productivity was as good as or better than the average for the industry (0.30 man-hour to turn out one airframe pound compared to the average of 0.47 man-hour). All in all, an impressive record—but this rate occurred after the Air Force needed it. Ford built some remarkably sophisticated jigs and fixtures, but out of 21,000 such pieces of production equipment, only about 11,000 were ever put to use. Willow Run took too long to bring on line, and the Ford engineers never came to grips with the constantly changing design details dictated by new tactics and strategy emerging from combat zones. "The realization that the B-24 design could never be frozen came hard to the Ford engineers," historian I.B. Holley wrote. "Only gradually did they begin to understand that design change was a perfectly normal attribute of military aircraft. . . ." Production of engines involved much less design change, and licensed auto manufacturers like Chevrolet, Ford, and Buick turned out more than a third of Pratt & Whitney's total production of 356,000 engines.[35]

In many respects, the development and production of the Boeing B-29 Superfortress epitomized the striking achievements of the aircraft industry in attempting to maintain a controlled pattern within a changing, kaleidoscopic production enterprise. During World War II, the U.S. produced some 3,895 versions of the Boeing B-29 Superfortress, a plane that was roughly twice the weight of a B-17 (120,000 pounds compared to 60,000 pounds). Its size and sophistication, including remotely controlled electronic gun systems and pressurized crew compartments, made its production record even more remarkable. At a cost of some $3 billion, it represented the most expensive aircraft program of the American war effort. Except for the atom bomb—delivered by these bombers in 1945—the B-29 was probably the largest and most complex weapon system of the war. The process of constructing cavernous new factory buildings, acquiring innumerable machine tools, training thousands of unskilled workers, and assuring proper fit of a myriad of components (some 40,540 separate parts) produced in various locations throughout the country required new standards of management. With buildings, tooling, and production components under way before the first experimental plane took off (in September 1942), the B-29 represented one of the first U.S. Air Force ventures to rely on "concurrency." "The B-29," wrote one admiring journalist in 1945, "is the most *organizational* plane ever built."[36]

The complexity and sheer size of the B-29 influenced its design and construction features. The main fuselage section emerged from designers' drawings as a straight cylinder 40 feet long. This permitted construction of an uncomplicated fuselage without weight and labor penalties of added structural reinforcements. Also, large sections of the fuselage panels all had the same measurements and fabrication characteristics, simplifying tooling and production costs. This inherent simplicity, despite its size, meant that the plane would be easier to build, especially by younger, less skilled workers. Boeing, along with Air Force managers, perfected the technique of fabricating large sub-assemblies at locations scattered around the United States, then shipping them to final assembly lines where less experienced workers could literally bolt everything together more quickly and with fewer mistakes.[37]

Boeing's responsibility for the B-29s was far from over when the planes rolled out of the factory door. Pilots and flight crews required extensive instruction in order to fly and operate the planes' myriad systems. Ground crews needed similar education for maintenance and repair of mechanical units, hydraulics, electrical connections, guns, bomb racks, miles of circuitry, engine systems, and a thousand other components. Like others in the war, when faced with an unprecedented complexity of systems and little time for formal instruction, Boeing executives found short cuts. Training films graphically displayed exact production procedures—and the consequences of a bungled job. The production of such films for hundreds of tasks, plus the writing, illustrating, printing, and distribution of a basic 2,000-page manual (in five carefully structured topical volumes) in itself became a small industry. Many separate systems required individual training and maintenance documents. The company prepared such publications along the model of widely read magazines like *Popular Science* and *Popular Mechanics*, with an easy-to-read text and airbrushed illustrations to focus attention on specific details of a procedure.[38]

The flood of aircraft orders, fed by tens of millions of dollars in contracts, raised the specter of profiteering. Congress wanted to keep competitive bids alive but acceded to the realities of accelerated production and the immense numbers of contracts involved. During 1940, procurement shifted from competitive procedures to the idea of cost-plus-fixed-fee contracting, so that such types represented 55 percent of Air Force contracts valued at $10 million or more. The fee structure allowed 4 percent of cost during the war years, and the profit margin on such agreements usually amounted to half that of fixed price contracts. Further legislation mandated compulsory renegotiation in cases that suggested excessive profits. By and large, the wartime profits of aviation firms seemed eminently fair and reasonable.[39]

Regarding production totals, the Undersecretary of War, Robert Patterson, appropriately capsulized the achievement with a paraphrase based on Churchill's famous quote about the RAF during the Battle of Britain. As Patterson said, "Never were so many provided with so much." Between 1939 and 1945, American manufacturers turned out 324,750 aircraft, beginning with an annual rate of 4,935 in 1939 and reaching peak production of 96,318 during 1944. Some 231,000 units were purchased by the U.S. Army Air Forces with the rest delivered to the U.S. Navy and to Allied units. Because the gross figures made no distinction between diminutive, two-place primary trainers and massive bombers, the comparison of airframe pounds procured by the Air Force offered an especially significant record: 20,279,000 pounds in 1939, 797,120,000 pounds at peak production in 1944, and a total of 2,089,436,000 from 1940 through 1945. The annual production of engines for Air Force and Navy planes went from 22,667 in 1940 to a total of 812,615 at war's end.[40]

The record of American military aircraft production during 1939 through 1944 also stood out in comparison to other major wartime powers:

(1939–44)	Japan	Germany	United Kingdom	United States
1939	4,467	8,295	7,940	2,141
1942	8,861	15,556	23,672	47,836
1944	28,180	39,807	24,461	96,318
Totals	68,057	111,787	117,479	257,645

SOURCE: Irving Brinton Holley, Jr., *Buying Aircraft: Materiel Procurement for the Army Air Forces* (Washington: Government Printing Office, 1964), 555.

From the mid-1930s through World War II, the aviation industry moved from a modest position to a secure niche in public awareness. The general aviation sector, with a wide range of aircraft types, established patterns of market development that carried into the postwar era. Airline transports of American design captured international headlines and achieved a high level of sophistication in the form of four-engine, pressurized models that promised to revolutionize commercial air transportation before World War II abruptly canceled opportunities in the civil market.

It should be remembered that the survival of many companies in the Depression era had much to do with exports, especially military sales. The military factor in rapidly accelerating production just before America entered World War II also rested heavily on orders from abroad. Throughout the era, innovations such as flush riveting, folding wings on naval planes, and flying qualities often came from individual producers. At the same time, the role of the NACA in expanding its research facilities and collaborating with manufacturers illustrated the fundamental significance of federal investment in advancing the state of the art. All of these factors, plus attention to infrastructure, appropriate funding, and workforce training, provided the means for impressive production records during the Second World War. It would be these same factors, and more, that would contribute to the aviation industry's evolution in the postwar era. During 1945, the Aeronautical Chamber of Commerce became the Aircraft Industries Association, reflecting aviation's rising status in the industrial sector.

CHAPTER 4
Cold War Responses 1945–1969

IN TERMS OF ORGANIZATION, technology, and industrial culture, American military aviation in the postwar era underwent fundamental changes. For one thing, scientific thinking and planning became embedded in an Air Force matured and sobered by the events of World War II. Toward the end of the conflict, the Air Force wanted to maintain the advanced scientific capability represented by the civil sector's expertise. It consequently allocated $10 million to Douglas Aircraft for the establishment of the RAND Corporation, so named from a contraction of the term "R and D," standing for research and development. The RAND venture carried out conceptual studies of aircraft engines, advanced aircraft designs, and rockets; and analyzed postwar military issues, the logistics of advanced weapon systems, and the applications of the fledgling computer science industry. RAND became an independent, nonprofit business in 1948, although it principally did its work through research contracts funded by the Air Force. Additionally, the Air Force gave increasing responsibilities to various commands within its own structure to conduct advanced scientific and technological hardware development.

High-speed flight research rapidly expanded, often in collaboration with the NACA. On October 14, 1947, Air Force Captain Charles E. Yeager piloted the rocket-powered Bell XS-1 to a speed of Mach 1.06, the first person to travel faster than the speed of sound. The postwar decades included a series of record flights by exotic rocket planes in which the military services and the NACA probed the mysteries of high-speed aerodynamics and consequent stresses on aircraft structures. This work became adapted to successive generations of advanced combat planes, whose more conventional jet engines nonetheless eventually provided power to push them past the speed of sound and up to Mach 2 or more. Unlike the "X" series of individual, experimental aircraft, the military planes entered service in the hundreds of thousands, creating new demands on industrial equipment and production systems. The United States Air Force, created as an independent service in 1947, reached maturity as an intercontinental military presence in an "Air Age World."

The threat of Cold War clashes generated sustained congressional spending, and the role of postwar airpower, while affecting both the Navy and the Army, made the Air Force into a major customer for high-tech industrial products, with repercussions at home and abroad. The costs of postwar military planes prompted manufacturers to exert considerable pressures on potential buyers overseas since foreign orders could keep unit costs down, make the product more affordable, generate greater sales, and yield higher profits. All of this involved high-level politics and a fair share of imbroglios. Meanwhile, costs continued to grow, driven by expensive electronics as well as complex designs and exotic materials required by modern combat expectations and impressive speeds. New styles of management, production agreements, and something called "commonality" in Air Force and Navy designs were implemented to keep budgets under control. Another concept, "concurrency," related to the process of acquiring production tools and support equipment for operations while a design was still on the drawing boards. The idea was to speed the acquisition process and reduce costs. In both cases, such goals often proved elusive.[1]

Workforce and Market Changes

During World War II, the suppliers of military aircraft liked the business provided by government contracts but disliked the federal regulations that came with it. For many months after 1945, airframe and engine contractors tried to find substantial sales in the less-regulated public sector, a market that proved to be limited. The Aircraft Industries Association predicted that several of its members were doomed to fail. The Truman administration proved to be a savior, promoting a resumption of defense spending in the interest of national security. This agenda, pushed by the government, not the defense corporations, forged the early links of what came to be known as the "military-industrial complex" and created new corporate giants in the Cold War era.

For some companies, readjustments after the war's end were painful but successful. For others, readjustments were hampered by ill-considered judgments or simply came too late. Boeing, Grumman, and Douglas, among others, managed to find viable areas of production after 1945, especially as major defense suppliers when the Cold War raised international tensions. Other wartime leaders, like Consolidated and Curtiss-Wright, experienced considerable corporate turmoil. Consolidated survived; Curtiss-Wright did not.

Consolidated, the firm responsible for the famous B-24 Liberator and other wartime planes entered the postwar era as part of the very large and diversified company Avco (Aviation Corporation). Production units included Lycoming engines, Stinson personal planes, Consolidated, and Vultee, manufacturer of various military aircraft. The latter had been added in 1943 and linked to Consolidated to form a production company eventually known as Convair. The postwar era brought inevitable downsizing and change; Stinson went to Piper Aircraft in 1948. At facilities in San Diego, Convair developed postwar transports known as Convair Liners, while the Fort Worth division produced the gigantic B-36 intercontinental bomber. In 1953, another conglomerate known as General Dynamics bought Convair, and this division became one of the major aerospace firms of the postwar era. Lycoming engines continued as part of the original Avco enterprise. Convair itself became the corporate umbrella for building air transports, advanced bombers and fighters, missiles, and electronic gear. Although a unit of a large defense conglomerate (General Dynamics produced tanks and submarines), Convair adapted itself to the postwar era. Its civil airliners achieved a modest success; its missiles and military planes won billions of defense contract dollars.

By the end of World War II, the Curtiss-Wright Corporation trailed only General Motors in terms of corporate size. In the immediate postwar era, Curtiss-Wright's rapid downsizing to peacetime levels often led to elimination of R&D projects in order to maintain profitable balance sheets. Talented engineers and design personnel went elsewhere, joining automotive and other industrial firms. The corporation failed to implement projects for postwar transports; its wartime reliance on P-40 production (a mid-1930s design) left it behind in advanced capabilities for the postwar era. The company's big, turbo-compound Wright piston engines kept reassuring profits flowing in the early years after 1945, but management again failed to respond to the advent of gas turbine engines. Belatedly, Wright secured licenses from Britain to produce the Armstrong-Siddeley Sapphire (the Wright J-65) and Bristol Olympus jet engines, but misguided efforts to redesign them meant delays and another failure to secure postwar markets. Corporate headquarters in downtown New York City insulated management from the research and development realities occurring at production facilities. As one frustrated corporate executive wrote some years later, "Top management control by bankers and lawyers, professionals dedicated only to making money, was responsible for both the rise and fall of Curtiss-Wright Corporation." By the mid-1950s, Curtiss-Wright had been bypassed by the aerospace industry. The company carried on as a supplier of industrial valves, mechanical components, and overhaul services—often for other aviation firms—but was never again identified as a major player in the aeronautical field.[2]

The postwar readjustments in America led to major changes in the workplace as salaries contracted and thousands of workers were laid off when big military contracts ended. Strikes and work stoppages flared up across the country when steelworkers walked out along with employees from Ford, General Electric, General Motors, and others. The aviation industry, unaccustomed to intense labor confrontations, soon became affected. Boeing's labor relations had been relatively smooth, but continuing layoffs there, plus the highly visible actions nationwide, inevitably led to problems. During 1946, as thousands of Boeing workers received severance notices, the controversial issue of seniority came to the forefront when supervisors started "bumping" less senior people below them. Other issues turned on Boeing's proposals to exempt a percentage of people, such as skilled senior workers, from the bargaining unit so as to transfer them between jobs or shifts in order to keep crucial assembly work in progress. Management-labor tensions continued to rise into 1948, as neither side seemed able to draft an acceptable work formula. Company lockouts and union strike actions marked the next two years, until an agreement was reached in 1950. The four-year labor problem at Boeing was reflected in varying degrees at other companies. All aviation employers realized that large-scale labor questions now constituted an additional factor in the complex business of building planes.[3]

In time, expanding markets and new products brought thousands of additional jobs to the aviation sector, making it one of the nation's major employers and a significant factor in the national economy. In addition to the military sector, the improvements in aerodynamics, avionics, and power plants generated a major shift in the air transport manufacturing industry as the jet age transformed airline travel. The product line of leading manufacturers in the general aviation sector also expanded, most dramatically in the form of high-performance twin-engine planes and executive jets that entered service during the mid-1960s.

During the postwar years, especially after Korea, the aircraft industry experienced four major market changes. Following a fast start and an early slump, the demand for general aviation aircraft took off during the 1950s as personal and business flying rapidly expanded. From 1955 to 1963, general aviation sales doubled, driven by rising demand and enhanced by a diversified catalog of aircraft equipped with much-improved radio and navigational aids. Although the total sales in this market represented about 3 percent to 5 percent of dollar sales for the entire industry, the general aviation aircraft produced in America represented approximately 90 percent of all such types (tens of thousands) operated around the world. As a result, they constituted a constant, pervasive reminder overseas of the American aviation industry, and provided an important first step of familiarity that led to purchases of larger civil and military aircraft from the United States.

A second major change in the civil market came when airlines in America and overseas began buying jet transports. Boeing's 707 and 720 jets, the Douglas DC-8 jet, and the Lockheed Electra turboprop accounted for a big share of sales figures between 1958 and 1964. Jet airliners produced by the United States overwhelmingly dominated the airline fleets of the world outside the Soviet bloc nations.

Sales of military aviation products continued to dominate the industry, and the advent of guided missiles brought a third dramatic change, this one in the military sector. During the early 1950s, development of guided missiles meant successively larger annual shares of the Department of Defense budgets, accounting for one-quarter of industry sales by 1958. Taking note of this phenomenon, journalists and commercial analysts began to use the term "aerospace industry" and the Wings Club of New York solicited members from the missile and space industries—a further acknowledgment of the space age. The next year, the Aircraft Industries Association became the Aerospace Industries Association of America, setting the aerospace stamp on the postwar era. Within the traditional aviation business, this phenomenon meant restructuring and reallocation of resources. For the emerging aerospace manufacturers, the product mix changed dramatically. In 1956, the airframe companies averaged 5.71 percent of their military business as missiles; in 1961 the share of missiles climbed to 44.35 percent.

Closely related in terms of federal largesse, the evolution of the civil space program during the sixties comprised the fourth major change in the industry. The principal effort was focused on a manned lunar landing before the end of the decade. The multibillion-dollar program created new research centers and production facilities across the nation, and also transformed the old NACA into a new NASA, with far more management tasks than its predecessor. In addition to various economic impacts, the space program was also interpreted as a metaphor symbolizing American technological prowess and global leadership.

By 1959, the aircraft, aerospace, and parts industry had emerged as the largest employer in the United States; conventional classification included four main subgroups represented by airframes (including missiles), aircraft engines and parts, aircraft propellers and parts, and other aircraft parts and equipment. The vagaries of defense spending and the national economy invariably meant changes in rank from time to time, so that the industry as a whole came in second to motor vehicles and equipment in 1961, when the latter posted 715,800 employees compared to 701,600 in aircraft. Nonetheless, the size of the aircraft and aerospace industry made it a leading feature of America's industrial picture during the 1950s and 1960s.[4]

Milestones: Jets, Swept Wings, and Computers

As technological signatures of the postwar era, jet engines and sweptback wings became hallmarks of rapid change. Both had roots in European developments.

In America, the idea of jet propulsion surfaced as early as 1923, when the NACA published a pessimistic paper on the subject by an engineer at the Bureau of Standards. Although some work continued on components and design at NACA Langley, these tentative projects lagged far behind European achievements. By the late 1930s, Frank Whittle in England and Hans von Ohain in Germany had developed full-sized, gas turbine power plants for aircraft. Germany built the first jet airplane, the Heinkel He 178, which made a flight in 1939.

In England, Whittle conducted his own research, leading to the first successful bench-test of a jet engine in 1937. Somewhat belatedly, the Ministry of Defense gave the support needed for further research, including the construction of an airplane to conduct flight tests. With the advent of World War II in Europe, all of this remained highly classified. During an official visit to Britain in April 1941, Hap Arnold, Chief of the U.S. Army Air Forces, was surprised—and irritated—to learn about this British turbojet plane, the Gloster E28/39. Clearly, no similarly advanced aviation project existed in the United States, particularly at the vaunted NACA. The British aircraft had already entered its final test phase and, in fact, made its first flight the following month. Still fearing a German invasion, the British were willing to share their turbojet technology with America. With principal engine manufacturers like Allison, Pratt & Whitney, and Wright already committed to urgent wartime contracts for piston engines, American authorities selected General Electric to work with the British. GE's continuing development of turbine systems associated with aircraft engine turbo-superchargers made it a reasonable choice. That September, an Air Force major, with a set of drawings manacled to his wrist, flew from London to Massachusetts, where General Electric went to work on an American copy of Whittle's turbojet. An engine, along with Whittle himself, followed. Development of the engine and design of America's first jet, the Bell XP-59, was so cloaked in secrecy that the NACA learned nothing about them until the summer of 1943. Subsequent American jet engines in the early postwar era owed much to the British designs, which powered early Air Force jet aircraft like the Lockheed P-80 and others.[5]

Less well known than Whittle, but significant nevertheless, were the contributions of a Russian Sikorsky employee who started the United States on the road toward swept-wing aircraft in the jet age. Like several other chapters in the story of advanced flight research, the story began in Europe, where an international conference on high-speed flight—the Volta Congress—met in Rome during October 1935. Among the participants was Adolf Busemann, a young German engineer, whose paper advocated swept-wing aircraft for high-speed flight. In 1942, designers for the Messerschmitt firm, builders of the remarkable ME-262 jet fighter, realized the potential of swept-wing aircraft and studied Busemann's paper more intently. Following promising wind tunnel tests, Messerschmitt had a swept-wing research plane under development, but the war ended before the plane was finished.

In the United States, progress toward swept-wing design proceeded independently of the Germans, although admittedly behind them. The American chapter of the swept-wing story originated with Michael Gluhareff, a graduate of the Imperial Military Engineering College in Russia during World War I. He fled the Russian revolution and gained aeronautical engineering experience in Scandinavia. Gluhareff arrived in the United States in 1924 and joined the company of his Russian compatriot, Igor Sikorsky. By 1935, he was chief of design for Sikorsky Aircraft and eventually became a major figure in developing the first practical helicopter.

In the meantime, Gluhareff became fascinated by the possibilities of low-aspect ratio tailless aircraft and built a series of flying models in the late 1930s. In a memo to Sikorsky in 1941, he described a possible pursuit-interceptor having a delta-shaped wing swept back at an angle of 56 degrees. The reason, he wrote, was to achieve "a considerable delay in the action (onset) of the compressibility effect. The general shape and form of the aircraft is, therefore, outstandingly adaptable for extremely high speeds." Eventually, a balsa model of Gluhareff's concept, along with some data, wound up on the desk of Robert T. Jones, an NACA Langley aerodynamicist known as a maverick. Studying Gluhareff's model, Jones soon realized that the lift and drag figures were based on outmoded calculations. Applying new theoretical ideas, Jones made a breakthrough and sent initial reports to NACA directors in early March 1945. In a few months, advancing American armies captured German scientists and test data that corroborated Jones's assumptions. Within two years, the first swept-wing American jet planes took to the air.[6]

In retrospect, NACA's preliminary studies with gas turbines and swept wings represented a mixed record. Although there were tentative steps made in both areas, initial analysis of gas turbines led to the conclusion that the supportive technology lagged too far behind the theory. In Britain, Whittle was told the same thing, but the dogged engineer slogged ahead until private and government support caught up. Hans von Ohain had a somewhat similar experience in Germany. The U.S. lacked such an individual champion to buck the entrenched bureaucracy, and American authorities seemed so concerned about security issues that a blanket of secrecy prevented the interchange of information that enlivened the British aviation establishment. The American record in swept wings was somewhat better; Robert Jones picked up the challenge and followed it through. Still, it was no comfort to find that German swept-wing technology was so far ahead at the end of the war. There were postwar administrative changes at NACA to respond to these issues. In any case, as historian Alex Roland noted in his study of the agency, its shortcomings "should not be allowed to mask its real and significant contributions to American aerial victory in World War II." Moreover, NACA's achievements in supersonic research and rapid transition into astronautics reflected a new vigor and momentum.[7]

The European influences remained strong. By the end of World War II, the U.S. was keenly interested in tapping Germany's scientific and technological capabilities in rocketry as well as aircraft. Through a program known as Project Paperclip, the Americans contacted dozens of German scientists, engineers, and technicians, offering them salaries and jobs in the United States. Most of them accepted, since the destruction of Germany offered slim prospects in the way of jobs or a stable future. Not all were happy in America. Salaries were not always adequate, many jobs were inferior positions, and families found it difficult to adjust to a different culture. The majority, however, remained, becoming American citizens. There were a number of key personnel who made highly significant contributions to postwar aeronautical research. Hans von Ohain, turbojet engineer, Adolf Busemann, aerodynamicist, and Alexander Lippisch, a pioneer in delta-wing designs, represented only a few. Project Paperclip also swept up the research team that had developed the remarkable V-2 missile, along with their leader, Wernher von Braun.

At Lewis Research Center, NACA's engine development organization, engineers studied overtime to catch up on jet propulsion research from Europe. A flood of NACA translations of German test programs crowded the shelves of Lewis's technical library, where they became basic references in the new field of gas turbine research. Expatriates from Italy and Germany, now based at Langley and at the Air Force technical development center at Wright Field, made regular pilgrimages to Lewis to assist their American colleagues in grappling with these new aspects of flight research. Lewis carefully noted programs of the British Air Ministry and essentially followed the British lead in allocating research funds to turbopropellers (50 percent), turbojets (25 percent), and piston engines (25 percent). To cope with continuing problems of how to cool turbine blades in the new turbojets, Lewis garnered another Paperclip figure, Ernst Eckert, who played a lead role at the NACA center from 1949 to 1952, where he laid basic foundations for research into the esoteric world of heat transfer.

Still, the British maintained a strong lead in jet engines into the early 1950s, with U.S. firms manufacturing their designs under license. The Lockheed P-80 used a J-33 engine, originally built in America by GE and then mass-produced by General Motors' Allison division, based on the de Havilland Goblin engine. Early centrifugal type engines gave way to axial flow types that enhanced operation at higher pressures and facilitated slimmer and more aerodynamic airframes. Grumman's U.S. Navy fighter, the F9F Panther, achieved success with versions of the Rolls-Royce Nene engine, delivered to Grumman by Pratt & Whitney as the J-42; later models of the Panther used the Rolls-Royce Tay/Pratt & Whitney J-48.

The British lead in turbojets began to slow in the mid-1950s, as U.S. defense expenditures resulted in advanced test facilities that promoted new avenues of research. Aggressive management at Pratt & Whitney also pushed development of improved axial flow engines (the double shaft, or two-spool type); industry-wide research in metallurgy, heat transfer, and various engine systems paid handsome dividends. Despite its late start, Pratt & Whitney pulled ahead of rivals like Wright, Allison, and Westinghouse in the turbojet field. By the late 1950s, Pratt & Whitney, along with pioneer General Electric, dominated the market for turbojets, although Allison continued to compete in the gas turbine field by developing a series of successful turboprop engines.[8]

Computers and electronics played an increasingly vital role in the postwar development of engines, airframes, and missiles. During World War II, the myriad calculations required for long-range gunnery at sea and other ballistics conundrums led to the development of primitive electronic computers to generate useful gunnery tables. After the war, aviation and aerospace firms forged solid links with the computer community. Early postwar computing equipment proved eccentric and erratic, features that many commercial firms—banks and insurance companies—found to be too troublesome and expensive to warrant their use. Aircraft engineering groups were less bothered by quirky computers, having grown accustomed to the vagaries of radar and aircraft electronic systems during the war. In the postwar era, the teething problems of high-speed, jet-propelled aircraft made the teething problems of computers seem less irritating. In any case, the daunting flood of aeronautical engineering problems, translated into ream after ream of data, made the computer an alluring harbinger of better days to come, and Department of Defense contracts offered the means to fund continuing improvements in computing equipment.

The widespread entry of computers and electronics into the American aviation and aerospace industry came largely through the actions taken by one of the pioneering aircraft firms in California. After the war, Northrop Corporation sponsored two different computer projects to enhance advanced research for aerodynamic designs and develop a compact guidance system for a winged cruise missile called the Snark. These early computer ventures never quite attained the sophistication or reliability Northrop wanted but looked very interesting for further development. Meanwhile, Northrop's experiences in the development and application of computers spawned a number of new companies that advanced the role of computerization in the aerospace industry and elsewhere. One of these was Computer Research Corporation (CRC), formed in 1950 by a dozen Northrop engineers who resigned in order to launch their own electronics research and consulting enterprise. CRC became successful, and several different groups from its staff spun off to organize their own firms. This splintering effect characterized much of the postwar computer business; one study later identified some 14 different companies as descendants of the original renegades from Northrop. CRC itself survived its own cloning long enough to be acquired by a much larger corporation, National Cash Register, that wanted to enter the burgeoning computer market.

During the 1950s, aircraft companies in Southern California became the drivers for the expanding computer industry. Late in the decade and through the 1960s, the evolution of digital circuits led to development of powerful, compact guidance systems, first for USAF and Navy rocket projects, then for NASA's program to achieve a manned lunar landing. The miniaturization process rested on several different engineering developments but became a major factor as a result of missile research. Microcircuits also became part of increasingly complex aircraft functions for weapon systems, navigation, radar, flight controls, communications, and other requirements. By 1957, electronic components represented as much as half the cost of a missile, with aircraft showing a similar trend. It was no accident that a large number of professionals with aircraft companies held degrees in electronics or computer engineering.

On a different scale, computers and associated electronic systems for data relay and displays evolved into a sophisticated $8 billion air defense system known as SAGE (Semi-Automatic Ground Environment). It was refined during the 1950s and became fully operational by 1963, remaining in service into the early 1980s; the IBM Corporation functioned as the chief contractor, with dozens of subcontractors involved over the life of the program. The SAGE program triggered a number of important technological legacies. Among the early SAGE descendants during the 1960s were new systems for air traffic control, airline reservations (the SABRE System of American Airlines), and new disciplines for computer programming. Based on SAGE research conducted at MIT, later benefits influenced simulation equipment, computer-assisted manufacturing, and computer graphics. By the 1970s, the pair of latter functions had become known as CAD/CAM, for computer-aided design/computer-assisted manufacturing, which allowed engineers to "draw" different component shapes on a screen, change them at will, and consider how they might interface with other structural elements. All the while, the CAD/CAM system stored all the necessary data associated with these different sequences to be recalled for final design specifications and drawings. Postwar electronic devices like these made possible the supersonic and nuclear-armed planes and missiles of the Cold War era.[9]

The Outlines of Postwar Combat Trends

Up until World War II, the distinctions for military designs remained fairly clear-cut, and planes were ordered as trainers, fighters, or bombers. The exigencies of fluid combat requirements during the war began to blur such distinctions. As postwar jet engines, electronics, and aerodynamics drove up costs for individual types, Air Force specifications became less individualized and more generalized. The concept of multimission aircraft became the norm, and some planes designed for a single role were refitted for added roles. Pilots of the Lockheed F-104 Starfighter, built as a high-speed interceptor, also found themselves in intensive training for low-level ground attacks. "Commonality" for both

Air Force and Navy military designs eventually led to American projects like the controversial F-111 swing-wing combat plane and the European Multi-Role Combat Aircraft, or MRCA, that became the Panavia Tornado of the 1970s.[10]

Helicopters played increasingly significant roles in the low-level environment, particularly in light of the U.S. Army's determination to develop and deploy its own close air support independent of Air Force planning. Capitalizing on the work of the brilliant Russian émigré Igor Sikorsky, America had developed a strong lead in helicopter technology. Sikorsky's VS-300, flown in 1939, set the pattern for modern helicopters with a large rotor blade atop the machine for lift and propulsion and a smaller tail rotor to counteract torque and help control the aircraft. Sikorsky's basic design entered limited production during World War II, and some early models were used in rescue operations in Burma before the war ended.

In the postwar years, Sikorsky concentrated on large, powerful helicopters used by the U.S. armed forces. The market for smaller, utility helicopters in military and civil operations was dominated at least for several years by Bell Aircraft. Bell's and Sikorsky's engineers, along with innovative development from firms led by Frank Piasecki, Stanley Hiller, and Charles Kaman, continuously improved helicopter design and engineering. Hughes Helicopter also organized during the 1950s as rotorcraft entered the U.S. military inventory and found extensive sales in foreign military service as well.

Lawrence Dale Bell, founder of Bell Aircraft, developed a keen interest in helicopters and began hiring a research and design team during World War II. In 1946, the small and unadorned Bell Model 47 received certification for commercial use from the Civil Aeronautics Administration, and the first commercial sale followed in the same year. With a need to expand its new helicopter operations, Bell decided on a Texas location. The wartime work in the Dallas–Fort Worth region had created a pool of suppliers and engineers, the climate was agreeable, and the economic considerations in terms of land, labor, buildings, and taxes were all favorable. By the end of 1951, Bell occupied several existing facilities, and went on to build a new multimillion-dollar facility in Hurst. Much of the plant's early production went to Korea, where Bell helicopters evacuated 25,000 wounded from the battlefield to lifesaving medical facilities. In the process, the ubiquitous little Bell Model 47 became an important component of hospital units in forward areas and was immortalized as the incoming chopper on the remarkably successful television program M*A*S*H*.

Frank Piasecki's original company, renamed as Vertol, developed a large-cabin helicopter with rotors at either end. With its elongated fuselage and upswept ends to house the helicopter's rotor gear, the H-21 "flying banana" proved useful in airlifting heavy loads as well as up to 20 combat troops. Along with a smaller H-25 version, Vertol sold some 700 helicopters through the 1950s. As interest in helicopter operations grew, Boeing decided to enter the field by way of acquisition and settled on Vertol. In 1960, Vertol's 2,300 workers became employees of Boeing. A year later, in recognition of the company's diversification in airplanes, missiles, and helicopters, the Boeing Company was announced, including the Boeing Vertol Company. At Boeing Vertol, aggressive development of Vertol prototypes continued, such as the CH-47 Chinook family, with production running well into the hundreds.

By this time, helicopters relied on gas turbine power plants, a trend pioneered by the Kaman Aircraft Corporation, which flew a single-turbine helicopter for the first time in 1951, followed by a twin-turbine installation in 1954. Concurrently, the U.S. Air Force sponsored work on compact gas turbines of light weight and high power for military service, and General Electric's T58, flight-tested in 1957, became the benchmark unit for subsequent gas turbine power plants. Helicopters and their engines had created yet another postwar market opportunity for the aerospace industry.[11]

At the same time, helicopters as well as larger aircraft like transports and bombers experienced modification and specialization for specific tasks, and the air arms of the United States and other major powers fielded a bewildering variety of such aircraft variants for electronic warfare, submarine patrol, weather reconnaissance, photo reconnaissance, and a broad spectrum of other duties. In the case of bombers and electronically enhanced patrol planes, the increasing role of computerized systems often determined the nature of the plane's mission and a new, official designation in the increasingly arcane world of electronic warfare.

The U.S. Air Force developed an awesome tradition of long-range strategic bombing in World War II, and this capability continued to influence American suppliers in the postwar era. The Convair B-36 represented the apotheosis of propeller-driven long-range strategic bombers. With its six engines in the pusher position across the wing trailing edges, eventually augmented by four turbojets in twin pods mounted outboard under the wings, this aerial colossus boasted true intercontinental range. But the transition to jet bombers came swiftly. At the end of World War II, the lode of German aerodynamic research reoriented the direction of design by American manufacturers. Even before the Germans signed the terms for unconditional surrender, special Allied teams in Project Paperclip and other designated units headed for German research centers to locate key personnel and load up valuable reports. Early reviews of research notes from these sources prompted North American and

Boeing to revise some of their conceptual designs and redraft them to incorporate swept wings. North American's plan for a straight-winged Navy jet appeared as a swept-wing fighter and evolved into the F-86 Sabre; Boeing's proposal to the Air Force reappeared as a radical swept-wing bomber, the B-47. The B-47 influenced Boeing's subsequent bomber designs and became the precursor for a long line of phenomenally successful jet airliners whose configuration became universally adopted.[12]

Boeing's swept-wing B-47 entered service in 1951, preceded by a handful of smaller jet bombers of very conventional design (the North American B-45, for example) that helped build a learning curve for the large numbers of increasingly complex jet bombers that followed. Still, production of the 600-MPH B-47 did not prepare manufacturers for all the difficulties inherent in designing and producing supersonic models requiring different approaches in power plant operation, electronics, and construction technique.

One of the most original designs came from Convair as the B-58 Hustler, the world's first supersonic strategic bomber. The plane was designed as an integrated weapons system, carrying fuel and weapons in an aerodynamic "mission pod" under the fuselage.

The B-58 emerged from a welter of bomber studies funded by the Air Force in the late 1940s and early 1950s. Beginning in 1953, the project received funding at $15 million to $20 million per year in order to generate a final design and plan the manufacturing process. The plane's unique design, speed, avionics, and bomb load (thermonuclear, biological, chemical, or conventional) made it more of a research and development program than a routine design effort leading to straightforward production. Once a final configuration had been established, unique construction features for the wing and other surfaces—a honeycomb design with a cellular core sandwiched between two thin, metal skin sections—created severe problems in tool design and fabrication technique. Late in 1954, the Air Force had to decide whether to cancel the program and lose its investment of $200 million or keep the program alive through flight trials scheduled in 1956 for a total of $500 million. The program continued; its partisans in the Air Force believed that the experience gained from bomber operations at supersonic speeds and the production lessons from the B-58 effort would pay off in the future. But the program's erratic progress and costs continued to make it highly controversial.

The plane finally took to the air late in 1956, and the Air Force subsequently placed a series of orders totaling 290 aircraft. Convair and some 4,700 vendors geared up for production. But budget constraints and shifting priorities led to deep cuts in B-58 procurement so that less than 120 planes were left on order. The total costs for this bomber force now stood at $3 billion, a figure that led to sensational news stories about the new plane that cost more than its weight in gold, calculated at $35 per troy ounce. Early in 1960, the production run was set at 116 bombers, each costing some $14 million, or 20 percent more than earlier estimates—three times the cost of a Boeing B-52 and 10 times the price of a B-47. Despite its dazzling speed and high-tech design (requiring expensive maintenance), the B-58 fell victim to its specialized design in a period of rapidly shifting military technology. As Convair discovered, winning the contract for a high-priced, glamorous performer like the B-58 did not automatically bring uncontested annual funding, extended production, nor the profits that normally attended a healthy production run. Still, with an active vendor list of 4,926 firms in 44 states, the production of the B-58 underscored the economic effects of such complex weapons systems in terms of its nationwide allocation of $3 billion in contracts.[13]

The racy Hustlers were replaced by the huge 200-ton, eight-engined Boeing B-52, capable of flying unrefueled for some 13,000 miles. During the late 1940s, the Cold War environment led military planners to consider an advanced intercontinental bomber. At Wright Field, the thinking coalesced on a design using turboprop engines and equipped for aerial refueling, although Boeing kept a turbojet bomber alive in its own design department. When Pratt & Whitney announced a 10,000-pound-thrust jet engine, Boeing promptly worked it into a new proposal to take to the Air Force during the fall of 1948. The Air Force seemed interested and, on the advice of one of its émigré German experts, suggested a plane with a more swept-back wing, like the B-47. Ed Wells, head of the visiting Boeing delegation, hesitated. The B-47 had been flying for less than a year, and the aerodynamic challenges of the swept-wing goliath now being proposed seemed unusually intimidating. Finally, he agreed.

It was Friday afternoon, and Wells told his Air Force counterpart to expect them back early Monday morning. In their hotel rooms, the Boeing team went to work—new drawings, a wing swept at 35 degrees, new calculations for performance, range, and so on—a whole new plane by Saturday. Then one group made a foray to buy some balsa wood, glue, paint, and carving tools for a model while a second group worked with a stenographer and printer to produce new documentation. By Sunday night they were finished. On Monday morning a bemused Air Force colonel received a totally revised proposal, along with a large-scale model of a new jet bomber called the B-52. During the plane's development, another German expert at Wright Field helped convince the Air Force that Boeing's wing design would yield the promised performance.[14]

The B-52 joined Air Force units in 1955 and remained in production for seven years, a total of 744 planes progressing from A through H models. Missions were usually planned for cruise speeds of more than 500 MPH, although the big bomber could reach more than 600 MPH. Its size gave it remarkable adaptability, making it economical to operate through several decades. More than 400 of the big bombers equipped the Strategic Air Command in 1974, and upgraded versions remained in service through the 1990s, the planes being older than the crews who flew in them.

The longevity of the B-52 in different roles (low-level, launch platform, electronic warfare) mirrored the virtues of successive electronic suites for flexibility in the continually shifting nuances of Cold War requirements. The changing dimensions of modern warfare meant the proliferation of specialized aircraft in almost infinite variety. The deployment of large numbers of Soviet submarines sparked varying countermeasures, including airborne patrols. The advent of enemy subs carrying nuclear missiles added urgency to the development of planes that could spend many hours aloft and carry a heavy complement of radar units, echo-ranging devices, magnetic and electronic detection gear, and communications equipment. Rather than remove bombers from service and rebuild them for patrol duties, the military services found it easier to acquire large civil transports and load them with electronic furnishings for early warning or countermeasures. Modified Super Constellations for the U.S. Air Force and Navy ranged far out to sea, covering critical ocean approaches to the continental United States. The plane's fuselage became packed with detection and tracking gear, machinery to cool it, a compact operations center, bunks for sleeping and relief crews, and spaces for 30 or more crew members to monitor and coordinate all this equipment.

The accommodation of electronic gear became an integral factor in designing new combat planes. Developed in the early 1950s, about the same time as the B-58, the Convair F-102 was one of the earliest fighter-interceptor planes designed and built from the start with its aerodynamic framework, engine, avionics, and armament considered as an organic whole—a weapons system. More than that, designers considered its maintenance, support equipment, test program, and repair manuals as integrated features of the plane. Tooling and production all proceeded concurrently, using an elaborate engineering management system to coordinate everything in progress. Enabling the scheme for work such as this engaged the efforts of mainline organizations like the NACA and the Air Force's Arnold Engineering Development Center at Tullahoma, Tennessee, and spilled out to embrace Navy centers, universities, and industry.[15]

The manufacturing of these milestone combat planes—the B-52, B-58, F-100, F-4, plus other Mach 1 and Mach 2 fighters and bombers entering service in the mid to late 1950s and 1960s—also demanded new levels of sophistication. Requirements for light-alloy forgings of unprecedented size led to special research conducted by Alcoa (for metal production) and Wyman-Gordon (machinery for component production), whose partnership in a heavy press program was funded by the Air Force for some $397 million. Through a contract from the Air Force, the Massachusetts Institute of Technology spent four years of intensive research (1952–56) in the development of numerically controlled machine tools. Another $180 million went to develop new types of advanced machine tools for production from companies like Cincinnati, Kearney & Trecker, Giddings & Lewis, Onsrud, and others. Military requirements contributed to a new industry dedicated to the production of 500–600 tons of titanium each month. Other projects led to compact, lightweight avionics as well as hydraulic systems able to operate under extremes of cold and heat. The British aviation expert Bill Gunston called the advances "prodigious . . . exceptional" and said they "transcended everything done earlier"; he pegged total costs at more than $2 billion.[16]

To achieve lightness and strength, many sections of aircraft structures were fabricated in the form of a sandwich, with very thin skins on the outside and various "honeycomb"-styled filler materials between. Lightweight panels found use in structural sections of planes like the B-58 and F-111 and soon appeared on civilian jet airliners, usually as ailerons and flaps. There was also considerable progress in the use of adhesives and bonding agents to reduce utilizations of rivets in order to save weight, to achieve very smooth surfaces, and to create the honeycomb/sandwich structures. During the 1950s and 1960s, additional techniques had to be developed for working with exotic new materials like sheet magnesium, titanium, and high-stress aluminum alloys. One new technique, devised in order to produce components from heat-resistant, high-strength materials, was electrical discharge machining (EDM), which used an electrical current to remove or cut materials. An electrolytic grinding machine was used in profiling honeycomb cores without burrs or remnants. Milling machines became increasingly automated and programmed for multiple tasks; Giddings and Lewis offered one such tool as a five-axis profiling machine. As an example of early applications of such trends, the F-100 used the first one-piece integrally stiffened skins on a fighter. Compared to the earlier F-86D, the F-100 also used six times as much titanium; during the mid-1950s, North American bought 80 percent of all titanium alloy used in the United States.

Absorbing the costs of acquiring new machine tools, exotic fabrication materials, and computing equipment represented only a few of the postwar adjustments demanded of aerospace manufacturers. These factors plus new ways of doing business created intricate management arrangements, especially for the spreading practice of subcontracting. During World War II, roughly 50 percent of major programs relied on subcontracts, a procedure that died out afterwards because of associated management costs and the desire of manufacturers to exercise more in-house control. The Korean War, 1950–53, precipitated an abrupt reversion to the practice, based on a sudden influx of orders that exceeded existing capacity and a deliberate federal agenda to encourage dispersal of industry as a matter of national security. Moreover, the government encouraged subcontracting in order to support small businesses and to generate employment in areas experiencing a labor surplus. Following Korea, ongoing Cold War concerns and new programs pushed subcontracts up to a level of 30–40 percent. Automotive and other non-aerospace manufacturers rarely received these subcontracts. More often than not, the prime contractors made their arrangements with other aeronautical firms that seemed more attuned to the higher levels of quality and reliability expected for aircraft and aerospace components.[17]

Even so, the rise and fall of specific programs often left many companies at risk when procurement cycles inevitably entered a downswing. At one time or another, nearly all the major aerospace firms ventured into diversified businesses as a hedge against the vagaries of military contracting. Grumman made canoes, boats, and truck bodies; Convair produced buses and household appliances. For the most part, these were not profitable activities. Companies that went into the related fields of missiles and aerospace electronic systems, like North American, Martin, and Convair, experienced better results. Martin eventually left the airplane business to focus on missiles and electronics, and many others eventually negotiated mergers.[18]

Throughout the postwar decade, aircraft costs continued to spiral higher, despite the perennial efforts of congressional committees and independent commissions to identify and control the causes. While subcontracting, with associated management costs, contributed to the issue, the complex mix of factors seemed a byzantine puzzle. Complicated tools and costly alloys represented genuinely higher expenses. The increasing complexity of the aircraft themselves played a part, even for a single-seat fighter plane. For example, the engineering effort for the North American F-86D, produced in 1951, just nine years after the P-51, required six times the number of drawings (6,500) and three times the hours of engineering time (1,132,000). Avionics comprised a growing share of aircraft costs, representing 25 percent of the cost of a plane and 35 percent for a missile during the late 1950s and early 1960s. The manufacturers were not always at fault either, when government proposals stipulated complicated designs that contributed to high weight and construction costs. In fact, many design engineers blanched at the government's specifications and their implicit associated expenses, turning out leaner and better designs instead.

One such designer was Douglas Aircraft's Edward Heinemann. In the early 1950s, the U.S. Navy issued requirements for a new multi-crew jet-powered attack plane to weigh some 45,000 pounds. Heinemann took the weight specification as a challenge. He believed needless equipment and crew members pushed weight-cost spirals. His dogged doctrine of simplicity resulted in the A-4 Skyhawk, which weighed 10,000 pounds empty and 17,000 pounds loaded.

Not all aircraft programs reflected Heinemann's success with the A-4 Skyhawk. Within the industry, costs often remained excessive, the result of high-priced executives, bloated staffs, and "gold-plated" hardware. At the same time, Congress imposed requirements for cost accounting and miscellaneous record keeping that added layers of bureaucrats and reams of paperwork to the federal system as well as the manufacturer's. Moreover, powerful individuals in Congress still indulged in log-rolling, pork-barrel politics that created further defense costs.[19]

Even with dedicated designers and engineers, enhanced sophistication of machine tools, and new skills in using exotic metals like titanium, the complexities and politics of the postwar years made research and development of advanced fighters a risky process.

Politics and Airpower: F-104 and F-111

For some major programs, marketing aircraft abroad became an essential element of corporate strategy. The story of the Lockheed F-104 Starfighter is a vivid example of the importance placed on overseas contracts. It also illustrates the kind of role played by talented designers like Kelly Johnson, who relied on a compact, closely knit group of people known as a project team to keep complex developmental engineering endeavors under control.

With a major in engineering, Kelly Johnson graduated from Michigan in 1932 but was originally turned down by Lockheed due to lack of experience. He went back to school, took a master's degree in engineering, and found a warmer reception at Lockheed on his second try. Early assignments included work on a variety of civil and military designs, and he served as chief research engineer for the P-38 of World War II. Eventually, Johnson presided over a collection of visionary engineers and technicians who referred to themselves as the "Skunk Works," after the secret location of a still for producing moonshine in the popular comic strip "Li'l Abner." They prided themselves on the development of a series of imaginative designs, characteristically turned out in minimum time. Their prototype for the first American production jet fighter, the P-80 Shooting Star, rolled out of the Skunk Works just 143 days after Lockheed got its directive to build the plane. The Skunk Works model, which predicated an isolated project team focused on a single goal, became legendary at Lockheed, especially in the context of developing the highly classified aircraft popularly labeled as "spy planes." These included the slow-flying U-2 reconnaissance plane, built with long wings like a sailplane to cruise at high altitudes, and the SR-71, built to fly even higher and carry out intelligence missions at Mach 3 speeds. The Skunk Works model served as a pattern for similar operations in the aerospace industry. In the 1980s, when NASA's Johnson Space Center began serious designs for its space station, it sequestered its own Skunk Works crew in an off-site office complex while they hammered out a baseline design.[20]

Following his work on the jet-propelled P-80, Kelly spent most of his time on combat planes. Early in the Korean War, Kelly made a tour of air bases in the theater of operations. The consistent remarks from U.S. fighter pilots who tangled with the MiG-15 stressed the need for a plane to fly higher and faster than the enemy. Back in the United States, Kelly and his design team went to work on small, lightweight fighters. Each successive plane embodied new innovations or broader use of advanced technologies in the quest for added performance and reliability. In the case of the F-104, wing construction relied more heavily on skin sections prepared from aluminum billets, precisely machine-routed to form integral stiffeners and tapered for requisite strength and minimum weight. The F-104 wing also used the first boundary layer control system on a production fighter plane. The arrangement used air bled from the engine and dispensed through dozens of tiny openings along the wing to smooth airflow and enhance lift for landing this very "hot" airplane. Titanium skin formed part of the aft fuselage where engine temperatures ran high.[21]

The initial design team for the F-104 Starfighter numbered about 50 people, reflecting Kelly's preference for small, cohesive groups like the Skunk Works. Sure of a production order after prototypes flew in 1954, Lockheed formed a special group of another 20 production engineers to map out manufacturing procedures to get the plane into active squadrons in the shortest time possible. When a design detail came out of the Skunk Works, it went directly to the production team, which proposed several different fabrication procedures. When all design drawings were complete, Kelly's entire group analyzed the proposals, selecting the

The Lockheed F-104 Starfighter was one of the early combat planes designed and built using the concept of concurrency. It was also one of the first significant multinational production ventures. The model shown here was delivered to the Danish Air Force. *Courtesy of Lockheed Corporation.*

individual manufacturing sequences that presented an integrated scheme having the least increases in design weight, drag, and cost. Kelly estimated that this procedure saved $10,000 to $12,000 for each plane manufactured. The concentration on "producibility" also contributed to a very successful transition to production overseas.

By 1958, NATO allies were scouting for a new supersonic fighter. With tacit support from U.S. government authorities, Lockheed seized on this golden opportunity. The U.S. Air Force had originally planned large orders for the F-104, but the plane's limited range and lack of space for expanded electronic suites put a halt to production. Lockheed was keen to utilize its surplus production capacity through foreign sales. West Germany appeared as the key, since they had a sizable and immediate need. Challenged by competing planes from England and France, as well as Grumman in the United States, Lockheed still felt that the F-104 offered performance advantages. Nonetheless, the company geared up for an unusually intensive sales campaign, for the foreign market alone carried a value of over $2.25 billion.

Lockheed won the contract and negotiated comprehensive co-production contracts with firms in Germany, the Netherlands, Italy, and Belgium for deliveries to several NATO partners. Licensed production also began in Canada and Japan. By the mid-1960s, production of the European Model F-104G in America and these partner nations exceeded 1,900 planes and represented a hefty source of income for Lockheed. Total production from all sources came to about 2,500 aircraft, with 740 built by Lockheed in California. At the time, this large overseas production constituted the most extensive international manufacturing effort in Europe since the close of World War II. Indeed, the F-104 production organization set a precedent for a global effort. Seven nations, from Europe to Japan, committed 21 major aircraft companies to the task, along with 31 big electronics firms. Most of the foreign work occurred in Europe, and Germany accounted for the lion's share of the effort, mustering 100,000 workers in 25 factories to turn out components for the airframe. GE's J79 engine was also produced in Europe for the Starfighter program, at Bavarian Motor Works (Koblenz, West Germany) and the Fabrique Nationale, in Brussels. At least six more companies took part in engine production, monitored by the American prime contractor, General Electric. Dozens of other German subcontractors supported the fabrication network.[22]

Lockheed, in its assiduous pursuit of contractors and support for the F-104G venture in Germany, entertained lavishly and doled out a fortune in bribes. Admittedly, gratuitous payments to politicians and various officials in order to secure overseas contracts did not represent a new gambit for either American or foreign manufacturers. At the same time, even those Europeans with extensive exposure to hotly contested international projects expressed amazement at the extent of Lockheed's offensive. Ronald Harker, dispatched from England to Bonn, Germany, to represent Rolls-Royce engines, recalled that "the battle for the German order had been ruthless." Nearly all of the representatives from competing firms stayed at Bonn's Königshof Hotel, where the "highly professional" Lockheed team, with top management and huge expense accounts, dominated the action. Harker also remembered that he and other company representatives nicknamed the hotel as the "Lock-heedshof."

After some years of rumors, the whole ugly story surfaced in the 1970s. The company's remarkable success in winning European contracts, it turned out, had much to do with high-level politics played by a royal agent—Prince Bernard of the Netherlands, who received a commission reported to be $1 million. Additionally, the company's own corporate report in 1975 admitted to $22 million paid out to assorted officials in Europe. To be sure, Lockheed was not the only major American contractor to have been guilty of questionable practices during the 1960s; Northrop became the center of several embarrassing episodes involving marketing efforts on behalf of its F-5 jet fighter to customers in the Middle East, especially Iran. One of the prime brokers named in the Northrop saga was Kermit Roosevelt, a former official of the CIA and Teddy Roosevelt's grandson.[23]

Not every major contractor conducted dubious marketing campaigns—and some didn't campaign at all. Grumman, perennial supplier of Navy combat planes, did not even join the Aerospace Industries Association (AIA) until the late 1950s. Part of this aloofness had to do with the attitudes of Leroy Grumman, who didn't like advertising and the "dirty" politics of marketing; part was due to attitudes of engineers at the Navy's Bureau of Aeronautics, who viewed contractor efforts to "sell" the Navy as unprofessional. Grumman finally joined the AIA in 1959 at the express wishes of the bureau in order to facilitate various Navy contracts that also involved Grumman.[24]

For Grumman, this innocent step gave scant preparation for the public firestorms to come when the company became embroiled in the TFX decision, better remembered as the F-111 controversy. The debate occurred in the context of public concern about the "military-industrial complex," a term used by President Eisenhower in his farewell address, delivered in 1961. He warned of the potential dangers implicit in "this conjunction of an immense military establishment and a large arms industry. . . ." Soon after John Fitzgerald Kennedy became President, his Secretary of Defense, Robert McNamara, not only moved to slash defense costs but also to establish civilian primacy in weapons procurement as opposed to traditional reliance on the selection and review boards dominated by the military services. In the process, he changed the relations between contractors and their counterpart military bureaus and revised the role of Congress as a shaper of defense policies and procurement. The debate over his procedures—and their results—appeared in innumerable newspaper editorials, magazine features, and books.[25]

McNamara also sought to make the Department of Defense (DOD) more efficient by revising contract agreements and streamlining the acquisition process, emphasizing cost effectiveness every step of the way. Cost plus fixed fee contracts were discouraged because DOD staff felt they created wasteful practices and offered no benefits for good performance. Instead, the Air Force and other services were switched to a fixed price contract, spiced with incentive bonuses for good performance. McNamara also eliminated construction and test flights of competing prototype aircraft, presuming to save money by selecting one contractor for a project based on a competition of design proposals. In the competition for the mammoth C-5A transport, won by Lockheed, three initial bidders generated 35 tons of paper proposals for DOD analysis—enough to fill 14 DC-3 transports. McNamara's staff also introduced the Total Package Procurement Concept, requiring the contractors to bid the entire costs of research, development, and complete production run of a plane or missile. The C-5A was bid under this new approach. Lockheed convinced DOD that it could efficiently extrapolate existing jet transport technology into its new design, even though the C-5A weighed nearly three times as much as Lockheed's C-141 jet transport. The C-5A's unique size created major development snarls, all outside the project's inclusive contract, leading to a continuing battle over cost overruns. Such multibillion-dollar debates, with national implications for employment and public taxes, came to involve Congress, as budget and oversight committees took on expanded roles.

Following the McNamara era (1960–68), the DOD retained considerable civil control of weapons acquisition but agreed to U.S. Air Force preferences for prototype testing ("fly before you buy"); efforts to reduce mountains of paperwork; and scrapping the total package concept in favor of cost reimbursement contracts that featured periodic budget reviews at specified milestones in research and development programs. In the meantime, the TFX controversy took shape as the constraints of the McNamara era and the push for cost effectiveness influenced policy.

Drawing on research and development from the 1950s, the Air Force continued to plan a new attack/combat plane equipped with a variable geometry wing—the "swing-wing" concept. With wings forward, it could economically fly long distances or negotiate very-low-level target approaches with the aid of terrain-avoidance radar. With wings swept back, the plane could make a supersonic dash into the combat area and out again. A sizable weapons bay allowed internal stowage of conventional or electronically guided ordnance, including nuclear weapons. Originally designated as a strike fighter, implying commensurately larger size, weight, and complexity, it was relieved of the word "strike" during its gestation, and it became Tactical Fighter Experimental, or TFX. This name change, implying fighter-like size and agility, undoubtedly contributed to the brouhaha that unfolded. At the same time, the U.S. Navy hoped to acquire a plane of somewhat similar combat profile, including a wing design to provide crucial lift during operations from carrier decks. McNamara and his advisors decided that these two programs presented an opportunity to cut costs and directed the services to coalesce their studies into one aircraft, the F-111.[26]

In 1962, the Pentagon named General Dynamics as the "prime" contractor and assigned Grumman as "associate contractor" to develop an F-111B version for the Navy. With the total budget estimated at $7 billion to $8 billion, the largest awarded by DOD at the time, considerable attention became attached to it. News stories implied that the powerful congressional delegation from Texas brokered the award to benefit the aircraft division of General Dynamics in Fort Worth. Amidst growing debate, the Navy, the Air Force, and both contractors became riddled with internal debate; budgets escalated; and the plane's increasing design weight made it too heavy to operate from aircraft carriers.

By 1965, the DOD released new directives for General Dynamics to complete the F-111 program for the Air Force while the Navy began a separate program for its own plane. Unlike the F-111, Grumman's F-14 was designed from the start for aircraft carrier operations and aerial combat. Two years later, Grumman won the contract of $7 billion for the F-14 two-place fighter, which also featured a variable geometry wing. Along the way, Grumman's corporate culture became thoroughly politicized. During a major scheme for diversification in the late 1960s, Grumman developed its big Gulfstream II corporate jet and decided to relocate its production from the company's home plant in Bethpage, New York, to leased facilities in Savannah, Georgia, a city represented by G. Elliott Hangers (member of the House Committee on Armed Services) and home state of Senator Richard Russell (chair of the Senate Committee on Armed Services).[27]

The controversies and misinformation that swirled around the F-111 tended to obscure its essentially average history in terms of time elapsed for research, development, and deployment. Commenting on the highly publicized delays and problems in meeting original performance specifications, one comprehensive analysis included the statement that "[t]hese results were certainly unfortunate, but they were not really unusual." The author of the study, based on an exhaustive analysis of Air Force and congressional documents and backed by illuminating comparisons with other Air Force programs (including the later F-15), asserted that "the F-111 program appears far more typical in its results than the critics of the program would admit."[28] The F-111's initial problematic combat record also led to misinterpretations. Abruptly withdrawn from combat in 1968, the F-111 was not only reworked to cure structural problems, but intensive training and realistic practice missions also led to effective procedures for operational maintenance and full appreciation of the plane's startling abilities for low-level, night attack missions.[29]

With a variable-geometry wing and sophisticated avionics for low-level bombing attacks, the General Dynamics F-111 began as a joint project with Grumman Corporation and became one of the most controversial aircraft programs of the 1960s. *Courtesy of General Dynamics.*

The Sidewinder

Not all aerospace hardware evolved from elaborate weapon proposals awarded by one of the services to a big corporate contractor. The story of the Sidewinder air-to-air missile illustrated how innovation could originate in a modest federal laboratory and result in an amazingly effective weapon. Additionally, Sidewinder's deployment represented a strong belief in many circles that aerial combat in the 1960s would be decided by planes armed with rockets, not guns.

Among air-to-air missiles, the Sidewinder eventually acquired an enviable reputation. It was cheap; required low maintenance; was reliable; was devastatingly accurate. Curiously, Sidewinder sprang from no Air Force or Navy specification, but emerged from a Skunk Works environment at the Naval Ordnance Test Station at a place called China Lake, 150 miles northeast of Los Angeles in the menacing valleys and gulches of the

Mojave Desert. The missile's principal creator was William McClean, a bespectacled, bow-tied graduate in physics (1939) from Cal Tech. Through World War II, he worked at the National Bureau of Standards, where he displayed a remarkable affinity for mechanical design. He specialized in proximity fuses and guidance systems; at war's end he joined the Naval Ordnance Test Station, set up in 1943 to perform R&D for the Navy.

Eventually, McClean found himself hard at work on air-to-air rockets. Development teams elsewhere had decided to equip their rockets with large, complex radar systems capable of working through cloud cover. McClean reasoned that most interceptions of hostile bombers would occur above 20,000 feet, where clouds were not common. Infrared detectors would be very compact, cheaper, and usable nearly all the time in a cloudless, rainless environment. But nobody wanted to underwrite this challenge to prevailing wisdom. Using odd dribbles of discretionary money and near fanatical loyalty from a small, informal team that characteristically worked overtime, the bargain-basement missile began to emerge. Everybody seemed to love the challenge of developing a weapon costing only a few hundred dollars, characterized by commonplace procedures in manufacturing and ease of storage used for arming planes on carriers at sea. McClean's missile crew took perverse delight in the weapon's homespun design. Based on talks with one of its builders, a writer summed it up as "about as complicated as a portable radio combined with a washing machine." Lacking funds for top-dollar electronics, which were notoriously finicky in any case, the missile's telemetry derived from over-the-counter electronics kits produced for amateurs.[30]

In 1951, the Bureau of Ordnance bestowed its first significant development installment of $3 million. About the same time, McClean's group officially named their creation Sidewinder, after the desert rattlesnake that sensed its prey by detecting an emission of heat. By 1953, aerial tests proved that the Sidewinder could locate and hit elusive targets. During the same period, the Air Force pressed ahead with a complex, radar-guided missile called the Falcon. Following a dramatic competition using live missiles, the Air Force enthusiastically adopted the Sidewinder. It became operational in 1956, continuing in service long after the Falcon was phased out in the late 1980s.

Production orders for Sidewinders went to Raytheon and Ford Aerospace as joint contractors, and Thiokol supplied a solid propellant motor. Over the years, various upgrades and modifications occurred (with increased cost), extending production beyond the year 2000. Generally speaking, Sidewinder was about 9.5 feet long, 5 inches in diameter, weighed 190 pounds, and flew at Mach 2 with a range of more than 10 miles. Its no-nonsense design made it adaptable for ground-to-air versions and use aboard ships too small to carry larger antiaircraft missiles; its enduring simplicity made it compatible with virtually any sort of airplane in all sorts of operational environments.

During Vietnam combat, Sidewinder made aces of several U.S. pilots. Exported or license-manufactured, it was used internationally, including in RAF combat against Argentine planes in the Falklands campaign of 1982. It was lethal, posting a single-shot kill probability of 80 percent or better. By comparison, the Phoenix missile, with a length of 13 feet, diameter of 15 inches, and weight of nearly 1,000 pounds, could use its radar system to destroy multiple targets in terrible weather, but its cost in 1981 came to $413,000 apiece, compared to the Sidewinder's $38,000. In short, Sidewinder showed how creativity and simplicity could produce an exceptionally cost effective product.[31]

European legacies like jet propulsion and swept-wing designs played a central role in postwar aviation technology. As the Cold War tensions increased, defense budgets helped build America's aerospace industry as a major force in the national economy. In two, fast-paced decades after the end of the Second World War, aircraft speeds reached Mach 2, based on major improvements in propulsion, design, and fabrication. Electronics and computers became essential tools of the design process as well as integral features of the new generation of combat planes.

The aggregate size and presumed influence of the aerospace industry and the military-industrial complex certainly appeared awesome. By 1969, the four largest defense contractors included two that were directly involved with aerospace: Lockheed Aircraft Corporation ($2 billion) and McDonnell Douglas Corporation ($1 billion). The other two leaders were heavily involved with aerospace electronics, missiles, and aircraft: General Electric Company ($1.6 billion) and General Dynamics Corporation ($1.2 billion). To be sure, aerospace firms sometimes played fast and loose in winning contracts—especially overseas—and rewarding top executives with high salaries. The news stories of cost overruns did little to burnish an often tarnished image. At the same time, the apparently endemic cost overruns characteristically occurred in programs pushing the state of the art, when technological complexity and severe operational environments precipitated unpredictable problems; in the interests of national security, the rush to achieve operational status meant that fundamental research, engineering studies, and tooling for production all proceeded at the same time. As one corporate executive remarked, the aerospace industry was the only one that contracted to literally invent complex machines. For all the furor about tens of millions of dollars in overruns, a federal study in 1969 revealed only modest profits in the defense industry. As a whole, manufacturing industries in the United States received an average profit of 8.7 cents on every dollar, compared to 4.2 cents on the dollar for defense contractors.[32]

Collectively, the industry's experience with increasingly advanced aircraft and weapons systems contributed to a growing infrastructure of considerable sophistication. Such capability in turn played a role in U.S. responsiveness to other dimensions of Cold War concerns: ballistic missiles and space exploration for national prestige.

CHAPTER 5
Rockets and Space
1926–1969

IN 1955, AMERICA, self-assured as a world leader in science and technology, announced its ambitious intention to launch a scientific satellite in 1957–58 during the international geophysical year (IGY), a joint research program entered into by scientific groups around the world. A year later, the Soviets announced that they would launch bigger, more productive satellites. Cloaked by polite, diplomatic language about scientific contributions, the two Cold War antagonists raced each other to put a payload in earth orbit.

On October 4, 1957, the Soviet Union won the race, rocketing the world's first artificial satellite into orbit around the earth. Sputnik weighed 184 pounds, three times the weight planned by the United States. Within a month, a second satellite, carrying a dog named Laika, went into orbit, and the Soviets smugly announced the total payload circling the earth came to four tons. Vanguard, the American project to orbit a satellite, continued to endure a series of failures. A stunned public wondered how America's vaunted technological knowledge had failed to beat the Communists. Following Sputnik, the *Washington Post* declared that "the Russians managed the whole operation in such a way as to make convincing propaganda that they are ahead of the United States not only in scientific, but also in military rocketry." A later editorial decided that "the crisis before the country is in many ways similar to that which it faced in 1939–40–41. . . ." Labor leader Walter Reuther proclaimed the Soviet feat "a bloodless Pearl Harbor."[1]

The Eisenhower administration patiently endured a wave of criticism, aware that the situation appeared more serious than it really was. Intelligence reports, plus photo reconnaissance missions conducted by the super-secret U-2 over Russia, revealed that the U.S. still maintained an effective balance of power; the "missile gap" charge by John F. Kennedy in the 1960 presidential race did not constitute a national security threat. But Eisenhower could not reveal any of this highly confidential information. In the meantime, Eisenhower's administration had already given its blessing to a variety of programs in the mid-1950s destined to establish a strong position for future American space efforts. The public's response to Sputnik provided a clear opportunity to increase funding for the national space program and associated technical expertise in a variety of fields.[2]

Alarmed by the dramatic space symbols of Soviet scientific prowess, the United States intensified a determined effort to "catch up to the Russians." A presidential commission faulted science education, resulting in a broad national effort to reform the nation's educational system from kindergarten through primary and secondary schools. Revised curricula bristled with added scientific and mathematical coursework. Colleges and universities packed big classes in sciences and engineering, as new incentives from government and corporate funds reshaped the direction of higher education. The wording of the National Defense Education Act caught the urgency of national security implicit in the legislative title.

Rapid American progress in subsequent aerospace programs reflected the nation's sense of urgency and the government's budgetary largesse. At the same time, just as in the case of Lindbergh's flight, a considerable wealth of bankable experience already existed, an amalgam of traditional Yankee habits of tinkering, creative amateurs, military developments, and European expertise. With funding and direction, American space ambitions soon realized success.[3]

Foundations of Rocket Technology

The origins of rocketry remain obscure, probably dating from the thirteenth century, when the use of black powder rockets is indicated by Chinese sources. By the eighteenth century, European armies deployed barrage rockets, and sailors in the nineteenth century used rockets to propel whaling harpoons as well as to shoot lifelines to stranded ships. All of these rockets relied on solid propellants like black powder; twentieth-century visionaries eagerly turned to liquid chemical propellants to provide both fuel and oxidizer for high thrust in the "airless" region of

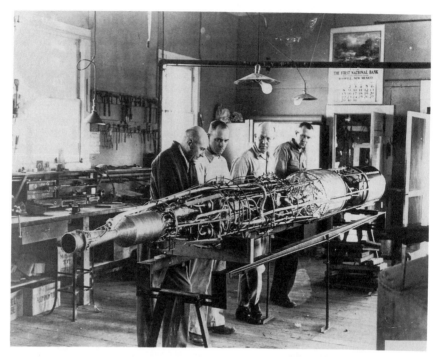

In 1940, Dr. Robert Goddard (left) and his assistants built this liquid-propellant rocket at Goddard's laboratory in the desert near Roswell, New Mexico. Although its design represented increasing sophistication in technology, American rocketry in World War II was paced by companies like Aerojet, based on the West Coast, and Reaction Motors Incorporated, based on the East Coast. *Courtesy of the National Aeronautics and Space Administration.*

space. Advances in metallurgy, chemistry, and the liquefaction of gases led to increasingly sophisticated and practical rocket designs. A number of Europeans, including Konstantin Tsiolskovsky in Russia and Hermann Oberth in Germany, analyzed promising lines of development, and Robert Goddard conducted experiments in America. But European politics and World War II meant that much of their work remained unappreciated until after 1945.

Professor Robert H. Goddard taught physics at Clark University, where he had received a doctorate in 1911. He occasionally lectured dubious students about space travel and rockets to the moon. In 1919, Goddard codified his lectures into a scholarly paper entitled "A Method of Attaining Extreme Altitude," which was published by the Smithsonian Institution and became a primer on propulsion estimates and design for future rocket scientists. More than many contemporary experimenters, Goddard spent considerable time designing and testing hardware with liquid oxygen oxidizers and gasoline fuels. In a field on his Aunt Effie's farm, not far from Auburn, Massachusetts, he fired prototype engines, whose screeching exhaust disturbed neighbors and startled their livestock. On March 16, 1926, one of Goddard's designs lifted off from Aunt Effie's pasture to climb 41 feet in 2.5 seconds—the first successful flight of a liquid propellant rocket. A shy and reticent man, the cynical news stories about his early theories and experiments made him even more guarded about his research. During the 1920s and 1930s, he filed many patents but had little to do with the increasingly significant work produced by rocket societies and university projects in America. During World War II, Goddard did research for the Navy; he died in 1945.[4]

Early group research began in the United States when the American Interplanetary Society organized in 1930, becoming the American Rocket Society in 1934. Concentrated in New York City, with members across the country, the society's early constituency was largely made up of science fiction writers and enthusiasts, including the popular historian Fletcher Pratt, who became the society's first librarian. Over time, more scientists and engineers joined, leading to the design, fabrication, and testing of experimental hardware. Their work drew heavily on papers published in Germany by the *Verein fur Raumschiffart* (VFR), or Society for Space Travel, and then began to exhibit more original thinking, including the development of regeneratively cooled engines.[5]

Another center of activity emerged on the West Coast. From his position as head of the Guggenheim Aeronautical Laboratory of the California Institute of Technology (GALCIT) at Pasadena, Theodore von Karman presided over an important evolution of rocket technology. Frank J. Malina began graduate work at GALCIT in 1934. At that time GALCIT studies in high-speed flight underscored the limits of propeller-driven aircraft; meteorological studies of cosmic rays at extremely high altitudes were being pondered; and Dr. Robert Millican, president of the California Institute of Technology (popularly known as Cal Tech), became a member of the Guggenheim Foundation's advisory committee on Robert Goddard's proposals for high-altitude sounding rockets. Out of this yeasty mixture came some of the first organized university research on space flight.

After a Pasadena newspaper ran a brief story about a GALCIT seminar on rocketry, two local residents showed up looking for assistance. John Parsons was a creative, self-trained chemist and Edward Forman was a skilled mechanic. They had experimented with a variety of black powder rockets but needed some practical as well as theoretical advice to proceed any further. Malina, now at work on his Ph.D. thesis, got von Karman's permission to do a study of rocket propulsion and flight performance, aided by Parsons and Forman, although neither had student or staff status at Cal Tech. The trio's work attracted increasing interest, and von Karman was able to allocate lab space for testing solid rockets as well as several small experimental rocket engines using liquid propellants.

In 1939, the government released special funds for a variety of flight-related work, including money for a rocket system to assist the takeoff of heavily loaded planes like bombers and flying boats. Jerome Hunsaker, at MIT, welcomed his contract to study deicing problems of airplanes, telling von Karman, "You can have the Buck Rogers job." Hunsaker's quip captured the tenor of most engineers' opinion about rocketry. The rocket work at Cal Tech masqueraded under the caption of Jet Assisted Takeoff (JATO) to avoid scorn, and the eventual organization of Cal Tech's Jet Propulsion Laboratory employed the same subterfuge. Despite skepticism, significant work occurred in the course of JATO development. As Frank Malina himself wrote in 1964, "The present-day American large solid-propellant engines are direct descendants of these first primitive . . . JATO rockets."[6]

When America began serious development of its capability for space flight in the postwar era, a considerable base of rocket manufacturers already existed. This specialized infrastructure evolved from early business ventures and government projects funded during World War II. By 1945, rapidly expanding Army and Navy budgets for production of combat rockets totaled about $1.3 billion. Federal momentum had picked up in the wake of correspondence during 1940 between C.N. Hickman, a sometime Goddard collaborator on military rocket research, and Frank B. Jewett, chief of the Bell Telephone Laboratories and president of the National Academy of Science. Hickman argued that military rocketry had potential but needed support. Jewett's network of powerful friends resulted in a division for rocket research within the National Defense Research Committee (NDRC). Most of the research involved solid propellant rocket weapons. Under the NDRC umbrella, striking progress occurred at the Allegheny Ballistics Laboratory in Maryland, using rocket motors to augment penetrating action of four-inch shells. Additionally, the NDRC supported work at the California Institute of Technology that led to successful motors for large aircraft rockets.[7]

By the time America became involved in the war, these activities had generated a cluster of active rocket development sites at a number of universities such as Cal Tech, George Washington, Wisconsin, Minnesota, and Duke, as well as a handful of industrial firms—Bell Telephone, Budd

Induction Heat Company, Hercules Powder Company, and others. By 1944, Hercules had set up a special operation, the Sunflower Ordnance Works, to supply dry powder propellants to the Soviet Union, which deployed massed rocket barrage units along the front.

Wartime demands forced rapid development of a variety of rocket weapons and the industrial facilities to produce them. With NDRC cooperation, Cal Tech perfected air-to-ground weapons during 1944. One of the best known was a large, five-inch, high-velocity aircraft rocket (HVAR). The Air Force irreverently dubbed it the Holy Moses, introducing it in combat during July 1944, shortly after the successful Normandy invasion. It proved to be highly effective, especially when fired from P-47 Thunderbolts against enemy tanks and other targets. Some of Cal Tech's own facilities were pressed into production to meet demand, turning out 100 HVAR units per day that were flown to Europe for use by eager Thunderbolt pilots. Other HVAR production batches went to the Pacific, where equally aggressive Marine pilots mounted them on Corsairs to strike Japanese strong points in tough battles across the Pacific islands. Strengthened Japanese fortifications survived HVAR attacks, so Cal Tech worked with NDRC and the Navy to devise a much larger weapon—one foot in diameter, more than 10 feet long, weighing 1,284 pounds with a 590-pound warhead. Known whimsically as Tiny Tim, it gave attack planes the power of a 12-inch gun and went into action during the landings on Okinawa.

At the same time, tens of thousands of barrage rockets of different types entered production for the Army and Navy. They were launched from Army trucks and tanks, and from a variety of Navy ships, including a special vessel built for bombardment rockets at the invasion of Okinawa. Individual combat troops in all theaters of the war fired thousands more small rockets from the shoulder-mounted weapon known as the bazooka. Toward the end of the war, the Army alone took delivery of $12.5 million worth of rockets per month, and both services relied on production of propellants, structures, and related elements from more than a thousand plants scattered across the nation.

In addition to this industrial network for solid propellant vehicles, work progressed on liquid-fueled rockets, although wartime production never reached the levels of their solid counterparts. A key development came at GALCIT when the Air Force funded parallel work on jet-assisted takeoff units based on liquid rocket engines. In 1944, Theodore von Karman and Frank Malina formally named the Jet Propulsion Laboratory to continue this line of research, launching the pioneering contributions of JPL. Although JPL continued to work on solids, the liquid rocket units represented one of its most important early contributions. Because the Air Force considered liquid oxygen too tricky to handle in regular operations, the rocket engineers at JPL spent much time in testing a long list of other chemical oxidizers.

In 1941, a solid JATO unit was successfully tested on a light plane (an Ercoupe) and a liquid propellant design worked successfully in flight tests with a Douglas A-20 twin-engine light bomber. A substantial order for JATO propellants followed, awarded to a brand new firm, the Aerojet Engineering Corporation. For use by flying boats, the Navy had its own liquid-fueled JATO research under way, carried out by Reaction Motors Incorporated; eventually the Navy settled on a simpler, solid fuel unit also produced in large numbers by Aerojet. In any case, military support created the opportunity for two new, pioneering liquid-fueled rocket companies: Aerojet and Reaction Motors.

Aerojet Engineering Corporation evolved from the cooperative efforts of early rocket enthusiasts who had found a place with the GALCIT group led by von Karman and Malina. Incorporation took place in 1942 with an initial capital investment by Andrew Haley, von Karman's attorney. Subsequently, in order to meet statutory capital requirements for wartime contractors, Haley made a proposal to one of his former clients, the General Tire and Rubber Company. This gambit worked, and the Aerojet-General Corporation resulted. After the war, Aerojet supplied liquid rocket engines for many vehicles, from the Aerobee sounding rocket to the main-stage and second-stage engines for the Titan intercontinental ballistic missile (ICBM). The company also diversified into guidance and control systems, liquefication units, nuclear research, and other products.[8]

While Aerojet prospered, another pioneering firm, Reaction Motors Incorporated (RMI), learned that becoming an early leader did not guarantee corporate success. Reaction Motors emerged from activities of the American Rocket Society in the 1930s. One member, James Wyld, solved a frustrating series of early engine test failures by circulating fuel through double walls of the engine thrust chamber. This regenerative cooling used the engine's propellants to prevent chamber walls from burning through. Along with three other members, including an electronics expert from International Business Machines Corporation, this quartet joined forces to form Reaction Motors late in 1941.

Located in North Arlington, New Jersey, RMI was a shoestring operation, consisting of a shop housed in half of the upper floor of a garage owned by a partner's brother-in-law. Wyld recalled that "we scarcely had two nickels to rub together, and our plant . . . was about as large as a rather spacious outhouse." They persuaded a skeptical official from the U.S. Navy to visit for a rocket demonstration, leading him to a remote spot in a local forest. Luckily, Wyld remembered, "it was a good run, and he departed quite enthusiastic." Immediately after Pearl Harbor, they won their Navy contract for liquid fuel JATO research and rented their own shop. Other work followed, such as an engine for the Navy's Gorgon guided bomb and other projects. By the early 1950s, Reaction had delivered more than 500 such engines.[9]

In the postwar era, Reaction Motors worked on a 1500–pound thrust chamber that burned alcohol and liquid oxygen. When the Air Force organized its program to crack the sound barrier, Reaction supplied the rocket plane's engine, designing a cluster of four chambers to produce 6,000 pounds of thrust. Variants of this power plant, the 6000C4, went on to power the Navy's Douglas Skyrocket as well as Bell's X-1 and its successors. The Cold War kept Reaction's engineers busy with increasingly powerful engines for ballistic missiles and cruise missiles; supersonic flight research led to the 60,000-pound thrust engine used in the X-15 rocket plane. Operating from a series of scattered facilities in New Jersey, Reaction Motors often maintained a precarious existence. Screeching rocket tests, accompanied by ominous vibrations and occasional explosions, had nearby neighborhoods up in arms. The company relocated to a vacant Navy ordnance supply depot, a move that resolved complaints but did little to improve financial health.

With 20 employees in 1943, 55 in 1945, and 473 by 1947, Reaction owed creditors $600,000 due largely to deficits from too many fixed price contracts. Fortunately, the Navy wanted to keep the company afloat and managed to get Laurance Rockefeller interested. From 1947 to 1948, Rockefeller arranged an infusion of some $300,000 to lift RMI from a financial grave. By 1952, the payroll listed some 640 employees, sales reached $4.5 million, and Reaction Motors earned $1.25 per share. The company caught the eye of other investors, and in 1953 Mathieson Chemical Corporation (Olin Mathieson formed the next year) bought a controlling interest. Continued growth took RMI to more than 1,600 employees in 1958 with sales of more than $24 million. Five years later, the Rockefeller-Mathieson interests sold out to another chemical firm, Thiokol Corporation, which had already developed its own solid motors and wanted to add Reaction Motors as its liquid fuel division.

As a division of Thiokol, the fortunes of RMI began to fade. Evidence suggests that its East Coast, experimental style did not fit well with Thiokol's conservative, Midwestern culture. As one veteran engineer from Reaction Motors put it, his company was "essentially an R&D mechanical engineering and design company which was purchased by a chemical company." Still, Reaction Motors left its mark in the course of these twilight years. Its XLR-99 engine powered the X-15 through a series of record flights. The Navy and Air Force contracted for 50,000 engines used in a rocket-boosted glide missile, the Bullpup. RMI often achieved success with technical components and products utilized in other programs. The company supplied vernier rockets used in seven Surveyor missions for soft landings on the moon. Big missiles like Atlas and Titan ICBMs used high-performance propellant valves developed under a special contract.

As a pioneering rocket manufacturing firm, Reaction Motors led the way in a number of trends, including turbine pump machinery, swivel engines for directional control, and self-cooling thrust chambers. The latter technology became standard in the industry during the 1950s. Aerojet used it for its Titan engines and Rocketdyne used it to fabricate Jupiter and Atlas engines, as well as engines for the Apollo-Saturn launch vehicles and the space shuttle. But Reaction Motor's basic product, small engines for tactical missiles, placed it in a very constricted market, especially as storable liquid propellants lost ground to improved performance from solid motors.

In contrast, Aerojet personnel worked on both solid and liquid propellants from the start. Andrew Haley, one of the Aerojet founders, implied that Aerojet succeeded because it developed both types of systems from the beginning, avoiding the one-dimensional engineering experience that seemed to hamper Reaction Motors. Some at RMI seemed to feel that its demise was due to inadequate funding from Thiokol; others made a strong case for an inability to merge two dissimilar corporate cultures. By 1971, Reaction Motors had dropped to 298 employees, net sales of $6.9 million, and a loss of $1.5 million. Thiokol looked for a new buyer, with no success. In 1972, Reaction Motors was liquidated.[10]

During the 1950s, smaller operations like Reaction Motors contended with increasing competition from the established aeronautical firms as well as major industrial giants like GE and others. Besides firing variants of the V-2 and producing nose cones for Atlas and Thor, GE also supplied guidance systems, accessories, and made engines such as the unit used on Vanguard's first stage. Sperry Corporation marketed guidance packages, and produced Sparrow Missiles for the Navy and Sergeant Missiles for the Army. Ryan, Northrop, Convair, Martin, North American, Lockheed, Douglas, and other aeronautical firms pursued a variety of rocketry-related activities, from supplying components to playing the role of prime contractor for ICBMs. Established chemical firms and other corporations with familiar names also contributed to a continuously unfolding infrastructure. Dimethyl hydrazide and hydrogen peroxide came from Food Machinery and Chemical Corporation; Pennsylvania Salt Manufacturing Company supplied fluorine; Union Carbide delivered various fuels and oxidizers. Additional participants in the missile business included Philco, Westinghouse, General Motors, Reynolds Metals, and a catalog of hundreds more, large and small.

Corporate organization charts reflected the growing significance of rocketry. North American had set up an Aerophysics Laboratory in 1945, leading to work on a supersonic ground-to-ground guided missile. During the 1950s, the laboratory's activities expanded to the point where different sections specialized in missile design and rocket propulsion. The Rocketdyne Division soon established itself as a leader in large, liquid propellant rocket engines; the Missile Development Division was eventually renamed the Space and Information Systems Division. A 1967 merger with Rockwell-Standard Corporation, a leading manufacturer of automotive assemblies and industrial components, gave rise to North American Rockwell, later renamed as Rockwell International. During the same year, two other major aerospace firms joined forces. As McDonnell Douglas, the new corporation established several divisions, with McDonnell Douglas Astronautics Company centered on the West Coast. (For convenience, the name Douglas is used in this chapter; merger details appear in Chapter 7.)

Universities like Cal Tech, MIT, and others played additional roles in the evolving rocket/missile complex, along with a growing number of government labs and test centers given over to space-related operations.[11]

Ballistic Missile Systems

The development of an industrial base for increasingly larger and more complex rockets made it possible to build long-range ballistic missiles carrying high explosive or nuclear warheads. A major source of expertise in this area came from German advances achieved during the Second World War and aggressively exploited by the United States during the postwar era. The principal source of this technological inheritance was represented by the German V-2 rocket, a weapon developed during the war by Wernher von Braun and a large R&D group at Peenemunde, a special research center on the shores of the Baltic Sea. The V-2, standing 46 feet high and capable of carrying a warhead at speeds of 3,600 MPH at an altitude of about 60 miles, had no counterpart in the Allied inventory.

In developing the V-2, von Braun and his team of rocket experts achieved signal breakthroughs in many areas of astronautical engineering, including inertial guidance systems; high-speed, high-pressure propellant pumps; and cryogenic engineering, particularly in the case of liquid oxygen as oxidizer. These technologies, and 132 of the key men responsible for them, were brought to America in 1945 and formed the nucleus of an unusually capable cadre of rocket designers and engineers. From 1945 to 1950, the "von Braun team" launched surplus V-2 rockets under the aegis of the U.S. Army Ordnance Corps at Fort Bliss, Texas, near the New Mexico border. In 1950, the Army transferred its rocket development work to the Redstone Arsenal in Huntsville, Alabama, where the Army Ballistic Missile Agency (ABMA) was officially organized in 1956 to expedite work on the Redstone rocket and expand several missile projects. In the process of collaborating with design and engineering colleagues, and with contractor personnel from General Electric, Chrysler, Rocketdyne, and other firms, the German experts within ABMA contributed much to American missile and space ventures.

Out of the experience of World War II, the alluring possibilities of space research through rocketry suddenly seemed close at hand. At war's end, the JATO work in liquid propellants led to a two-stage Army ballistic missile called the "WAC Corporal." The U.S. Navy had its own program of sounding rockets, including a vehicle known as Viking, built by the Martin Company. By the 1950s, an impressive stable of sounding rockets had come into being. Another line of development focused on high-speed cruise missiles and ballistic rockets developed by the Air Force. Cruise missile research left a significant legacy in terms of high-speed aerodynamics, inertial guidance systems, and other operational lessons. Among the most valuable legacies were the Navaho's booster rockets, since these engines found their way into several ballistic missile programs. Produced by Rocketdyne, improved versions of the Navajo engines went into the Army's Redstone and Jupiter tactical missiles, as well as longer-range ballistic weapons like the Air Force Thor and Atlas rockets.[12]

In the meantime, the policy decisions and management structure shaping a mature strategic missile program had evolved from events during the late 1940s. At the end of World War II, General H.H. Arnold convened his top scientific advisors for a review of the conflict's scientific and technological developments as an indicator of future trends to be followed. Theodore von Karman chaired the review, which later appeared in 1945 as a 14-volume publication entitled *Toward New Horizons*. Acknowledging the technical brilliance of the V-1 and V-2 programs, the conclusions gave far more weight to the concept of "air supremacy" in the context of jet fighters and bombers. The report supported work on air-breathing cruise missiles, but advocates of long-range ballistic rockets ran into a thicket of skepticism. Critics—with understandable reasons—raised a number of embarrassing questions about accuracy, reliable propulsion, and how nuclear warheads might survive the fiery, punishing reentry into the earth's atmosphere. Proponents had to agree that these issues were serious sticking points, but they argued that the need to find solutions, in the interests of national security, justified an intensive program of development.

Meanwhile the Air Force busied itself with a rather extensive program to develop workable missiles, with no fewer than 26 programs under way in 1946. Eventually, technical and budgetary considerations led to a focus on three projects for intercontinental weapons. Northrop had a contract for the cruise missile known as the Snark, designed for remotely controlled guidance. The Snark never proved satisfactory, although it certainly broadened the base of developmental experience. North American's big jet-propelled Navajo long-range cruise missile had the first inertial guidance system, and its rocket booster engines became adapted for the majority of ballistic missiles eventually fielded by the Air Force. However, it became a victim of skyrocketing costs and major development hassles. The Atlas, under contract with Convair, was a pure rocket but went through several lives between 1946 and the early 1950s, when Cold War tensions gave it crucial momentum.

By the early 1950s, the promise of smaller, lighter atomic warheads stirred ballistic missile partisans to action once again. Also, continuing reports of Soviet work in ballistic missiles gave a sharp sense of urgency to the debate. In 1953, officials in the new Eisenhower administration set up a Strategic Missiles Evaluation Committee. Headed by Dr. John von Neumann of the Institute for Advanced Studies at Princeton, this blue-ribbon academic group became known as the "Teapot Committee," in honor of von Neumann's penchant for endless pots of tea during the committee's deliberations. Its report, submitted early in 1954, stressed the need for formation of a high-priority program for ICBMs.[13]

In short order, Brigadier General Bernard Schriever received special authority to head up a crash program within the Air Research and Development Command (ARDC). Trained as an engineer at Texas A&M, Schriever eventually joined the Air Force, was commissioned, and received a master's in aeronautical engineering at Stanford in 1942. He served in the Pacific during the war, and then did considerable service at the Pentagon as a specialist in planning and development. The ARDC set up a special entity, the Western Development Division, headed by Schriever and operated out of a former schoolhouse in Inglewood, California. To avoid attention, Schriever and other personnel wore civilian clothes. All this belied the power that Schriever had, since he operated completely outside normal development channels. Schriever did not even report to his putative superior officers in the ARDC, but directly to the Secretary of the Air Force. If there were still issues to be resolved, he was free to go to the Secretary of Defense, or, if necessary, to Eisenhower himself. Having decided to develop an ICBM, the U.S. intended to lose little time. When Schriever built his staff, he was given carte blanche to recruit any individual he wanted, either military or civilian; his Air Force headhunter whisked around the country in a B-25 bomber to press home the priority of the ICBM program. This sort of management clout, plus a totally pliable Congress, characterized the unique nature of the Atlas project, the early cornerstone of the ICBM program.

There were other managerial features that made the ICBM business different from past Air Force operations and that formalized patterns that continued. Reliance on a neutral organization to provide a coldly objective technical assessment throughout the venture was one such feature. This entity would supply studies on program planning, management, and operations, especially in the area of integrating numerous strands of cutting-edge technologies. The group would not only work with the contractors but also assist in evaluating contractor performance. The Air Force selected Ramo-Wooldridge Corporation for this task. This company merged with Thompson Products Company in 1958 to form Thompson-Ramo-Wooldridge (TRW), an event that intensified charges from other contractors about conflict of interest, since TRW now had subsidiaries that advised the Air Force on the one hand, and contracted for hardware on the other. Even though the Air Force had required legal guarantees from TRW to avoid this issue, the need for a new arrangement was obvious. In 1960, the Air Force established a separate, non-profit entity, the Aerospace Corporation, "to engage in, assist and contribute to the support of scientific activities and projects . . . for the United States Government."[14]

Meanwhile, the Thor and Jupiter intermediate range ballistic missiles (IRBMs) were sited overseas. Thor, with Douglas as prime contractor, had been placed by the Air Force in operational service in Britain in 1958, with 60 missiles deployed there through 1963. The Air Force also dispatched more than 60 Jupiter IRBMs, built by Chrysler Corporation, to Italy and Turkey between 1958 and 1963.[15]

Once the Air Force became convinced that an ICBM was crucial to national security, Schriever made decisive plans to guarantee an operational missile. Borrowing from the experience of the Manhattan Project, he opted to follow "parallel development," with a separate associate contractor for each major system. This approach offered insurance against the failure of one contractor to deliver, stimulated competition for rapid development, and promoted expansion of research and development within the comparatively narrow industrial base in America capable of missile production. Further, some parallel systems might be interchangeable among other missiles to be developed. In the case of Atlas, Aerojet General received a second source contract for engines, with Rocketdyne as the first source. When the Air Force started its IRBM plan, the Douglas Thor incorporated some subsystems originally developed for ICBM vehicles. Finally, Schriever and the Air Force concluded that a totally separate, vehicle program was needed to assure an operational missile system, and decided on an alternate two-stage ICBM design. As a result, the Martin Corporation received a contract to produce the two-stage Titan I, using many components and associate contractors that had been initiated as parallel suppliers for Atlas. There were a number of setbacks in the Atlas development program, including some fiery explosions in flight and some difficulties in working out guidance programs. Success was ultimately achieved, however; Atlas became operational by mid-1959, and Air Force personnel installed them at missile sites across the United States. Additional silos were also prepared for the separate launch vehicle design, the Titan, which became operational with the Strategic Air Command in 1962. The Atlas and the Titan were the first two missile systems to come out of the ICBM program; both had the capacity, when launched from the United States, to hit any target inside the Soviet Union.

The underground sites provided a comforting degree of safety and security, but loading liquid propellants at the last minute was a tricky, time-consuming process. The Atlas, with more than 40,000 parts, kept maintenance and operational crews at high levels of stress. Pentagon mandarins wanted a less complex missile with a faster reaction time. In the meantime, chemical research during the late 1950s produced increasingly sophisticated solid propellants and more effective systems to control the thrust vector of the rocket motor. This led to the Boeing Minuteman I of the mid-1960s, with rapid improvements for succeeding weapons known as Minuteman II and Minuteman III. Later versions carried decoys and chaff to confuse enemy defenses, and mounted multiple independent reentry vehicle (MIRV) warheads. The first missiles became operational in 1962, and eventually some 1,000 of them were stored in hardened silos sunk 80 feet below ground level in wheat fields and cow pastures scattered through seven states west of the Mississippi.

Although the three-stage Minuteman series was manufactured by the Boeing Company, it carried the imprimatur of "weapon system integrator," for the complex missile series engaged a variety of specialized subcontractors and suppliers. Reflecting the Pentagon's earlier management strategy for Atlas and Titan, technical management became the responsibility of TRW Systems Group. The first-stage engine for all three versions came from Thiokol Chemical, and Aerojet-General supplied the second-stage engines. Third-stage engines came from either Aerojet or from Hercules Incorporated. The guidance and control system packages were developed by the Autonetics Division of North American Rockwell; Sylvania Electronics supplied ground communications; contracts for the reentry vehicle went to two manufacturers, Avco Corporation and the General Electric Company.[16]

By 1965, the Air Force possessed a mature ballistic missile system and industrial infrastructure. From a force of fewer than 50 Atlas and Titan I rockets in 1960, crews could launch 54 Titan II and 1,000 Minuteman missiles just five years later. (The Titan II missiles, despite their age and drawbacks as early-generation liquid propellant rockets, remained on alert until 1987; the MX, or Peacekeeper, went on duty in 1986.) All this progress did not come cheaply, with at least $17 billion expended to cover R&D, production, and the cost of building missile sites. In the process, as an Air Force history stated, "the industry portion of the ICBM program was gigantic." Subcontracting had become an accepted practice in the postwar aircraft industry, but the ICBM effort "elevated subcontracting to a grand scale." As early as 1955, the Atlas program listed 56 major contractors, with 150 majors on the roster two years later as the ICBM plan expanded. At the end of the 1950s, the ICBM and IRBM combined efforts involved 2,000 contractors in missile production and construction activities, with 40,000 men and women at work. The Air Force looked back on the undertaking with obvious pride. Writing in 1964, service historian Ernest Schwiebert boasted, "This composite program far exceeded, both in complexity and magnitude, the earlier Manhattan project."

The institutional culture of the Air Force experienced a radical shift under the impact of the ICBM program, and in 1958 the service was presenting itself as a full-fledged "aerospace power" with an arsenal of aircraft, missiles, and satellites. The service's annual budgets reflected this major change. In fiscal year 1964, the Air Force allocated $3.5 billion for aircraft and somewhat over $2 billion for missiles. Although the totals varied occasionally from one year to the next, the ratio remained fairly constant through 1989. A network of early warning satellites, radar-equipped ships and planes, and a string of land-based radars to detect hostile aircraft and missiles became part of the multibillion dollar Pentagon budget. The U.S. Navy also developed its own missile capability.

The various land-based missiles and manned bombers of the Strategic Air Command comprised two elements of a three-way National Security concept called TRIAD, rounded out by sea-based missiles carried in nuclear-powered submarines. Because air bases and missile silos represented fixed, known sites, defense planners remained concerned that they might be destroyed by enemy missiles, thus reducing America's capabilities for retaliatory strikes. To maintain a believable deterrence capability, the Pentagon began to develop the U.S. Navy's Fleet Ballistic Missile System. Beginning with the first operational cruise of the USS *George Washington* in 1960 through the USS *Will Rogers* in 1967, the Navy put 41 missile submarines at sea in its initial deployment of the Polaris system. This enterprise embraced a number of new development, production, and management programs.[17]

The early subs normally carried a complement of 16 solid-fuel Polaris missiles built by Lockheed Missiles and Space as prime contractor. The Polaris represented a unique system not only in terms of undersea launch but also in terms of missile navigation. Land-based ICBMs and IRBMs headed for their destinations from a known point, simplifying their guidance programs for trajectory. The process for Polaris used an inertial navigation system, set to a sub's known position as it set out on patrol. All variations in direction and position underwater were constantly fed into the sub's own computers, which updated separate target trajectories for each of the missiles.

In order to finish a product as complex as a guidance system, a multitude of groups and individuals needed to interact successfully. Within the Strategic Systems Program Office of the U.S. Navy, separate bureaus held responsibility for the missile guidance systems and for the navigation system of the sub to launch it. Civil organizations, such as the Charles Stark Draper Laboratory of MIT and the Autonetics group of North American, designed and produced these systems.

Inertial guidance and navigation systems emerged from the work of Draper himself as head of the MIT Instrumentation Laboratory (as distinct from the Draper Laboratory at MIT); Autonetics; German scientists of the Wernher von Braun team at NASA; and others, including Northrop Corporation. Both social and political issues played a role in contributions made by the various players. Draper's personal knowledge of dozens of well-placed military officers and bureaucrats who had worked in MIT's projects during World War II and who had taken degrees at MIT undoubtedly smoothed his way.

Polaris was followed by the Poseidon in the late 1960s. Soon, advanced Soviet antisub warfare systems pushed Pentagon planners toward new submarines and an improved fleet ballistic missile capability dubbed the Trident. The new sub and its missiles of greater range allowed them to operate within an ocean area at least four times the size of prior missile subs. Each Trident-class vessel carried 24 missiles and their overall efficiency meant that early plans called for only 10 submarines to replace all the Polaris types. Costs were still high, estimated at $5.1 billion for building the submarines, plus more than $7.6 billion for missile development and construction of port facilities.[18]

To maintain this level of funding and compete with the Air Force for defense appropriations, the Navy, like its rival, paid careful attention to politics and public relations. The Polaris team, particularly the Special Projects Office, left no stone unturned in the process of promoting their weapon. To placate the Air Force, Navy reports invariably stressed Polaris attacks on tactical rather than strategic targets, even though the missile's accuracy did not justify that distinction. Factories and laboratories with Polaris contracts flew special flags, not unlike the "E" flags used during World War II. The Special Projects Office, including its chief, Admiral William F. Raborn, Jr., blanketed congressional offices with reports and traveled constantly to technical conferences and business meetings to spread the word. Contractors were prodded to mount their own public relations blitzes, nationally and locally. Rumor had it that the Special Projects Office wielded a bigger public relations budget than the Navy's own Public Information Office but kept it hidden in innocuous accounts.

In addition to its reputation for aggressive self-promotion, the Polaris program became known for generating a new style of management for complex weapon systems being developed under tight deadlines. The Fleet Ballistic Missile System program, of which Polaris was a part, relied heavily on "Program Evaluation Review Technique," or PERT. This management tool had three principal characteristics: estimates of time required to complete specific tasks as submitted from personnel involved in such tasks; incorporation of mathematical formulas for achieving milestone events; identification of "critical path" sequences expected to take the longest times to achieve. Even though the origins of the PERT idea seem to go back to Chrysler's work with the Navy on the Jupiter program, it became identified with Polaris. First made public in 1958, PERT became a hit, hailed as the forerunner of a new era of space-age management techniques and more important an innovation than Polaris itself. In retrospect, PERT received mixed reviews. For some companies, PERT did not fit existing research and development structure. PERT never delivered the magical results that Pentagon managers ascribed to it, although some successful features of the program, such as the idea of visibility in tracking subsystems progress lived on. In the long run, one analyst concluded that management success in Polaris owed much to capable people, robust budgets, and a favorable political environment. But the most significant factors rested on the development team's opportunity to apply basic management principles without interference, and the team's own confidence in its abilities to do the job. "The supply of such people," the analyst cautioned, "is likely to be short."[19]

Space Flight and Lunar Missions

The ballistic missile programs, with details cloaked by the strictures of national security, nonetheless stirred considerable public interest in rocketry and the prospect of space travel. Astrophysicists effectively used military launch vehicles for scientific observations at extreme altitudes, and reports of these activities appeared not only in the technical literature, but in newspapers and periodicals, as well. The Army's Wernher von Braun, with his good looks, continental charm, and appealing German accent, had a knack for explaining the intricacies of astronautics with disarming clarity. Although nuclear-tipped missiles were ominous and the physics of space flight continued to intimidate large segments of the general public, the concepts of rocketry and space travel were becoming familiar—and interesting—to many postwar Americans.

For the general public, one of the important sources that helped legitimize space exploration was a series of articles that ran between 1952 and 1954 in *Colliers*, one of the most widely read weekly magazines of the postwar era. With support from von Braun, and a series of eye-catching, full-color artists' concepts of spaceships and interplanetary voyages, the *Colliers* series reached 12 million to 15 million readers with each issue. The magazine's success prompted a pair of books drawn from the articles and generated a considerable interest in news stories, radio, and television coverage. In the process, von Braun began to emerge as a nationally recognized figure. During 1954 and 1955, Walt Disney developed a trio of television specials along themes introduced by *Colliers*, and the audiences for these animated shows ran into the tens of millions. President Eisenhower called Disney for copies of the specials, which he apparently sent to the Pentagon. Additional periodicals and educational materials helped build public support for space programs, especially in the wake of Sputnik. All of these factors seem to have been part of the series of events that led to President Eisenhower's endorsement of the International Geophysical Year and an American satellite.[20]

Upstaged by the Soviet artificial satellite in 1957, American policy makers scrambled to organize an effective response to what appeared as a serious challenge to the nation's scientific and technical capabilities, to say nothing of global leadership in the Cold War era. When Viking, the abortive American rocket for the IGY satellite, chalked up continuing failures, the Eisenhower administration turned to the Army's team of rocket experts led by von Braun. In a crash program, the Army Ballistic Missile Agency (ABMA) modified a Jupiter C rocket with upper stages that it had been using for warhead reentry studies, developed a payload with Cal Tech's JPL scientists, and put America's Explorer I into orbit on January 31, 1958. To expedite future American efforts in space exploration, Eisenhower also agreed to a reorganization of the NACA, which had already begun preliminary studies for more ambitious missions in space. On July 29, 1958, the National Aeronautics and Space Administration replaced the original NACA.

The late 1950s and early 1960s were a time of bold planning, rapid expansion, and occasional chaos as new space centers evolved to conduct planning and management. NASA's growing bureaucracy included the original NACA centers of Langley, Lewis, and Ames, all of which continued their aeronautical research as primary activities, adding relevant aspects of astronautical work as the space program evolved. NASA also recognized the need for specialized centers to plan and manage a mature program of space exploration and manned spaceflight. Goddard Space Flight Center at Beltsville, Maryland, held primary responsibility for unmanned spacecraft. Marshall Space Flight Center originated in 1960, comprised of personnel and facilities from ABMA, and managed the development of the Saturn family of launch vehicles, with later roles in hardware for Skylab, the space shuttle, and other programs. Johnson Space Center began as the Manned Spacecraft Center (MSC) in 1961, sited at Clear Lake, between Houston and Galveston, Texas. For nearly 12 years, as MSC, this unit had full responsibility for selection and training of astronauts, design and development of manned spacecraft, and management of manned space flights. Kennedy Space Center was officially named a week after President's Kennedy's assassination in 1963. Located on the Atlantic coast at Cape Canaveral, Florida, this site dated to the 1950s as a government rocket launch facility. Kennedy functioned as NASA's center of launch activities for manned as well as unmanned vehicles. These and other centers generated regional commercial and economic growth, but none perhaps as much as the string of NASA sites in the American South. From Houston (Manned Spacecraft Center) to New Orleans (Michoud Assembly Facility and Mississippi Test Facility) to Huntsville (Marshall Space Flight Center) to Cape Canaveral (Kennedy Space Center), NASA locations represented a new "fertile crescent" of advanced research and engineering. The spinoff of their presence was seen in the scientific and technical activities of neighboring universities and public schools, regional business, medical service, and urban growth. By the late 1960s, they had also become major tourist attractions.

Following Sputnik, the national security community in America considered space in terms of military and intelligence purposes such as weather forecasting, communications, navigation, and especially surveillance. Although there were lively debates about using achievements in space exploration to polish the nation's technological prestige, the Eisenhower administration disliked that idea. The Kennedy administration felt differently. In the spring of 1961, following Yuri Gagarin's triumphant orbital flight in the Vostok space craft, Kennedy's advisors saw the Soviet success as a direct challenge to America's global leadership. For Kennedy, space exploration became the metaphor of twentieth-century achievement. In the context of the Cold War struggle, the United States needed to surpass the apparent Soviet leadership in technological capability and political influence. The abortive American attempt to support an overthrow of the Soviet-supported regime of Fidel Castro in Cuba—the Bay of Pigs invasion—was not a principal reason for the final decision to support a lunar landing, but it certainly constituted one of the goads for Kennedy to take action to counter Soviet presence in world affairs.

He took this position to the nation during an address to a joint session of Congress convened on May 25, 1961. In his speech, Kennedy made the challenge explicit, charging the United States to take on "a great new American enterprise . . . a clearly leading role in space achievement, which in many ways may hold the key to our future on earth." He boldly declared: "I believe we should go to the moon," and then set the end of the decade as time limit for achieving this national goal.

The Soviets took Kennedy seriously. In secret, they formulated their own program to beat America to the moon in the 1960s. The nations of Western Europe, still encumbered by the devastation of World War II, could only watch from the sidelines as the two superpowers geared up to outperform each other in space. At the same time, prescient European scientists and politicians took the first steps to create a cooperative program to build a European space infrastructure. But the early years of the space age were dominated by the Soviets and the Americans.[21]

America's manned lunar landing enterprise was named the Apollo program. Overnight, to the traditional NACA/NASA role of research and development were added the daunting tasks of executor for a nationwide production effort and manager of global operations for tracking and communications. To manage the complex industrial and federal sources of Apollo's development and to solidify its own operational base, NASA imported a number of Air Force officers with missile and other industrial management experience. There were other new executives as well, often imported from outside the civil service. James E. Webb was appointed administrator in 1961 after a notable career in both industry and government, having served as Under Secretary of State as well as director of the Budget Bureau.

The von Braun group of German émigrés was not the only foreign team to join the American space effort. In the early 1960s more than two dozen Canadians from the AVRO firm joined NASA. The group was principally composed of British subjects who had come to Canada during the 1950s to work on a new combat plane known as the Arrow. The plane featured a number of advanced design techniques and unusually sophisticated telemetry instruments for real-time flight data transmission. In 1960, the Canadian government abruptly canceled the plane about the same time that NASA was looking for experienced aeronautical engineers for the early test phases of Apollo development. As recently appointed project engineer for a NACA/NASA Space Task Group, Robert Gilruth had heard of the AVRO research and felt he needed the group's unique research expertise. When Gilruth hired them en masse, they represented about 20 percent of STG's early roster of 160 personnel. After contributing to the Apollo effort, they stayed in the United States and took up citizenship; many of them rose to responsible positions in the field centers as well as at NASA headquarters and with various contractors.

Clearly, the Apollo program dominated NASA's funding during the 1960s as space claimed center stage. In 1960, the total NASA budget came to $523 million, with $100,000 for space flight studies. In 1962, NASA's budget was $1.6 billion, with $160 million for space flight. NASA's biggest money year came in 1966, when it was allocated $4.5 billion, of which $2.9 billion went to Apollo. By 1969, the respective totals were $3.1 billion and $2 billion.[22]

Reflecting on the legacy of Apollo, James Webb, the NASA administrator from 1961–68, cited a study from the Graduate School of Business at Columbia University. The most significant legacy, he said, came from NASA's style of management for advanced system design. As the authors of the Columbia analysis put it, the problem was "[g]etting an organizationally complex structure, involving a great variety of people doing a great variety of things in many separate locations, to do what you want, when you want it. . . ." Webb noted that the goal of a lunar landing within the decade "was clear and unambiguous," although NASA had to integrate several rapidly developing technologies that in turn evolved from rapidly forming scientific information. With Apollo at its peak, NASA's civil service force climbed to 33,000 at NASA centers, augmented by 390,000 contractor personnel and 10,000 people in university research. By the time of the first lunar landing in 1969, total personnel had dropped to 190,000, and by 1974 it had fallen again, to 126,000.

Atlas and Titan missiles from the USAF inventory supplied launch vehicles for incremental manned space flight experience in the Mercury and Gemini series of missions in earth orbit. Unlike previous aerospace efforts, in which manufacturers turned out hundreds of aircraft or missiles, Apollo ordered its principal vehicles in minuscule numbers: 10 Saturn I; 11 Saturn IB; 15 Saturn V. Less than a dozen lunar modules made space flights and only six made landings on the lunar surface. The launch vehicles, with multiple stages and up to five clustered engines, came closest to the image of serial production, and their costs stemmed principally from the exhaustive research and development programs that were necessary to ensure the utmost reliability in manned space missions. The total R&D and production costs for the Saturn rockets alone came to $9.3 billion dollars.

Building toward the ultimate goal of a manned lunar landing, NASA carried out programs such as Mercury and Gemini to gain experience in space flight technology and manned missions in earth orbit. Existing hardware like the Atlas and Titan ICBM missiles could be adapted as launch vehicles for these early operations, but there was no similar hardware for manned space capsules or the launch vehicles to carry out lunar missions.[23]

Not everything always went according to plan. Prototype hardware, designed for operational requirements where no precedents existed, periodically failed, and valuable time was consumed while engineers worked to pinpoint failures and devise a remedy. The most cruel tragedy occurred in 1967, when an Apollo command module caught fire during ground tests, killing three astronauts trapped inside. Investigation of the Apollo capsule fire uncovered a disconcerting gaggle of documentation failures. The investigation board discovered that it could not piece together a precise, detailed configuration of the spacecraft from available documents; details of fabrication procedures were incomplete; and a comprehensive list of all materials in the capsule was impossible. Rushing to get things done, NASA and its contractors had become sloppy in tracking changes and analyzing their interaction. The upshot was a more rigorous protocol for control. A new hard-nosed Configuration Control Board (CCB), chaired by George Low as NASA's spacecraft czar, took charge at headquarters. The board's 11 members included key headquarters engineers, key managers from Manned Spacecraft Center in Houston, and leading designers from North American and Grumman. The CCB convened every Friday at noon for no-holds-barred discussions that often ran late into the night.

The Grumman member represented that company's contract to design and build the lunar module (LM). Carrying two astronauts, the LM's descent engine had a throttle to allow a controlled, soft landing on the lunar surface. Like many other Apollo components, the LM ran into severe cost overruns, developed weight problems, and lagged behind schedule. Managers from NASA dispatched aggressive teams to Grumman (and other contractors) to analyze problems and recommend solutions. Generally, tough discussion sessions between NASA and contractors resulted in mutually agreeable remedies to revise procedure and get programs back on track. For instance, following one such interlude at General Electric, the supplier of checkout systems for Apollo, the GE manager wrote to Apollo Program Director Sam Phillips about the tough sessions with Phillips's emissary:

> As you might well guess he beat the living h—- out of us, . . . spurring us on to more effective utilization of our previously mammoth efforts. Despite the bruises, we feel that we are a far more effective organization now as a result of his leadership.

On another occasion, NASA managers brokered an exchange of technical information for the benefit of one company's troubled engine system. From Grumman, Bell Aerosystems held the contract for the LM ascent engine. Bell eventually solved early difficulties in welding and fabrication, but firing instability involving the fuel injector seemed an intractable problem. NASA informed Grumman that it had funded a separate injector design at Rocketdyne that proved successful after a 12-month development program. Reluctant to force another engine change so late in the LM development, NASA coordinated a team of specialists from both companies, even though they were competitors, to test Bell engines fitted with the Rocketdyne injectors. By the spring of 1968, NASA decided to use this hybrid, and the space agency's public relations literature listed both companies as the suppliers for Grumman's LM descent engine.[24]

In parallel with the evolution of spacecraft, NASA and its contractors began to develop a new family of launch vehicles to boost astronauts away from earth's gravity for the first time, en route to an historic landing on the lunar surface. The Saturn family of launch vehicles were direct descendants of a technological heritage that stemmed from the development of long-range ballistic missiles and other advances in rocket hardware after 1945. The Saturn vehicles were unique because they included the first rocket boosters designed and built specifically for manned launches; because of their remarkable size, nearly everything associated with them had to be rebuilt or scaled up to unprecedented proportions. Even with a solid technological base to start from, it was the process of scaling up that so often created problems and costly delays in the Saturn program.

The Douglas Corporation's Thor, a standard USAF intermediate range ballistic missile of the era, had a diameter of 8 feet and stood 65 feet high. Powered by a single rocket engine of approximately 165,000 pounds of thrust, Thor's liftoff came to about 110,000 lbs. The Saturn V boasted a diameter of 33 feet and towered 363 feet above its launch pad. Powered by a cluster of five engines, the first stage of the three-stage Saturn V generated a total thrust of 7.5 million pounds, with a liftoff weight of more than 6.1 million pounds. Trying to translate these awesome statistics into comprehensible terms, press releases frequently

In 1964, technicians make final inspection of a first stage for the Saturn I booster, a precursor of the Saturn V launch vehicles used for manned lunar missions. Assembled at a NASA facility not far from New Orleans, Chrysler produced the propellant tanks and fabricated the finished booster; Rocketdyne (a subsidiary of North American, later Rockwell) supplied the cluster of eight engines. *Courtesy of the National Aeronautics and Space Administration.*

noted that the Saturn V was as high as a 36-story building and weighed more than a good-sized Navy destroyer. Douglas, the first contractor to build Saturn's upper stages to contain liquid hydrogen (LH_2) as fuel, also had to cope for the first time with the problems of producing large, unitary propellant tanks. Unlike the complex, clustered tankage of the Saturn I first stage, the Douglas S-IV upper stages were single components, but their dimensions created a host of new problems in fabricating and production. Special, oversized jigs and welding fixtures had to be designed and built. New techniques of fabrication, metal bonding, and insulation had to be developed, and static test stands had to be constructed to unprecedented dimensions. The challenges of designing and producing the S-IV stage, in short, reflected many of the complexities faced by each of the stage manufacturers of the Saturn program.

Inevitably, costs increased, even though NASA maintained very close scrutiny of its contractors.[25]

Inheritors of a strong Army tradition of in-house R&D and manufacturing, the von Braun team at Marshall Space Flight Center was reluctant to let their contractors do all of the work. The initial five S-IC first stage boosters were produced by MSFC before Boeing took over as prime contractor. This effort was not made solely to accommodate the egos of MSFC's highly trained work force, but to maintain skills to back up contractors when they got in trouble and to provide NASA with a yardstick in monitoring performance and cost curves for hardware produced by the contractors. By and large, this approach proved effective and was one factor in keeping the Saturn program on schedule and within reasonable cost. The S-II second-stage contractor was North American Rockwell. At one point, the S-II stage became the problem child of the Saturn V enterprise, threatening to upset the entire booster program. The situation became so critical that some top-level managers at MSFC muttered darkly about the possibility of taking the S-II stage contract away from North American and giving it to another contractor. Before this dire step was taken, North American vigorously overhauled its managerial organization and appointed some new executives. Manufacturing techniques, especially welding technology, were refined with the aid of technical assistance from MSFC, and the rocky progress of the S-II program finally smoothed out.

In this instance, as in other cases, impassioned exchanges occurred between managers from NASA and their counterparts from the contractors. At the same time, individuals from both sides emphasized a sense of partnership—a genuine government/contractor team—that held together in order to achieve the historic objective of human exploration on the moon. A number of interactive management techniques kept team members in touch, and kept both progress and problems visible. To produce the Saturn family of launch vehicles, the process of tracking the seemingly overwhelming number of individual parts, components, and subassemblies for each engine and each stage called for unusually stringent management procedures. The Saturn Program Office at MSFC, which had major responsibility for this task, placed a good deal of emphasis on what it called the program control center, a room crammed with constantly updated progress charts, eventually duplicated by identical control centers at each major contractor plant and other NASA centers. Equipped with microphones and projection screens, these control centers facilitated executive meetings through simultaneous talks and visual presentations linking different locations across the country. Based on techniques originally developed during the Polaris program some years earlier, NASA's refinement of the idea thoroughly impressed U.S. Navy managers who toured the program control center at Marshall. "I used to think the ones we had in the Polaris program were good," said one of them, "but this puts us to shame." Regularly visited by touring management groups from America and foreign organizations like the Royal Air Force, the control centers became the model for scores of similar arrangements in the future.[26]

Eventually, the progress charts in NASA's control centers listed major elements of the Apollo/Saturn vehicles arriving for launch at Cape Kennedy. Beginning in 1967, two unmanned launches of the Saturn V and three manned missions over a two-year period qualified the hardware and systems designed for a manned lunar landing. Finally, on July 16, 1969, Neil Armstrong, Edwin Aldrin, and Michael Collins lifted off aboard Apollo 11. Four days later, Armstrong and Aldrin guided the lunar excursion module to a successful landing on the surface of the moon, where they carried out experiments and collected lunar samples. They rejoined Collins in orbit the next day and safely splashed down in the Pacific on July 24 for recovery by a team from the U.S. Navy carrier USS *Hornet*.

Jolted into action by the Soviet Union's Sputnik, the U.S. government funded comprehensive initiatives to develop new competence in space flight. Fortunately, the prewar years included useful work by enthusiastic amateurs and university research groups. This background enabled American industry to put solid propulsion combat rockets into mass production during World War II and set the stage for rapid postwar development. The arrival of Wernher von Braun and the inheritance of Germany's wartime research proved to be immensely valuable. During the 1950s and 1960s, American industry plunged into national defense programs that yielded impressive results in terms of IRBM, ICBM, and submarine missile deployment.

The technical achievements and managerial techniques of the military programs represented an invaluable resource for America's Apollo program to achieve a manned landing on the moon. In 1958, the NACA evolved into NASA. The agency's traditional role of research continued, but its resources were increasingly geared to space exploration, especially management of the hugely expensive lunar landing undertaking. International influences continued with the expansion of the "von Braun team" influence at Marshall Space Flight Center and the addition of groups like the AVRO Canadian cadre. Several new NASA centers brought visible changes to their locations, including the major impact of the South's "fertile crescent." During the 1960s, NASA and its contractors worked out admirably effective partnerships as the aerospace industry developed and launched space flight hardware for the historic journey of humans from earth to a different celestial body.

During the triumphal journey of the Apollo 11 and its crew, massive press coverage included live television broadcasts beamed to 33 countries on six continents, reaching 25 million viewers in the United States alone.[27] This dramatic triumph of astronautics certainly deserved all the attention it received, although progress in civil aeronautics also experienced significant change. Milestones in civil aviation and air travel after 1945, if not as dramatic in singular events, nonetheless amounted to revolutionary increases in the number of people who traveled by air and the speeds at which they flew.

CHAPTER 6
Airliners and Light Planes 1945–1969

IN ADDITION TO THE civil airliners that had been conscripted for wartime duties, hundreds of now-idle military transports were suddenly available at war's end for conversion to meet commercial airline needs. These included twin-engine Douglas C-47 and Curtiss C-46 transports in addition to much larger four-engine aircraft like the Douglas C-54 (DC-4 Skymaster) and Lockheed C-69 (L-049 Constellation). The military surplus equipment soon gave way to new planes as competition for air travelers intensified. Lockheed and Douglas effectively used their early designs for large, four-engine transports in the process of developing respective families of airlines—Douglas improving its basic DC-4 and Lockheed refining its L-049 design. The latter marked an important evolutionary milestone as the first pressurized air transport to enter mass production. For long-distance routes, pressurized transports were clearly preferable, and Douglas lost little time in refining its DC-4 into the pressurized DC-6. Boeing reentered the airliner market with a transport called the Stratocruiser. All of these pressurized airliners, boasting high operational ceilings and transatlantic range, meant enhanced passenger comfort and predictable scheduling.

By the mid-1950s, Douglas and Lockheed responded to the worldwide growth of air travel by developing stretched versions of these planes

(DC-7 variants and "Super Connie" variants) capable of carrying more than 100 passengers for 4,000 miles or more. When jet airliners began to enter service during the late 1950s, the revolution of air travel was already underway; jet transports accelerated the trend and spread it more rapidly around the globe.

The general aviation sector also experienced a dramatic expansion, recovering from the overoptimism that helped create a severe downturn in the early postwar years. Manufacturers offered a remarkable variety of general aviation models with a wide range of capabilities. The development of high-performance singles represented one notable trend, along with the introduction of pressurized twins powered by turbopropeller engines. During the 1960s, aggressive new entrepreneurs like Bill Lear created a new market for business jets. At the opposite end of the speed scale, helicopters established a small but promising market in civil aviation. Reviewing the impressive number of different aircraft designs that found buyers during the years between 1945 and the early 1970s, and considering the functional characteristics of helicopters, single-engine personal planes, utility aircraft, twin-turboprop business planes, and speedy corporate jets, the early postwar decades were a renaissance for the general aviation industry.

Changes in nomenclature accentuated this dramatic shift. After World War II, the Aircraft Industries Association of America, the leading trade association for aircraft manufacturers, recognized the promising status of general aviation by creating the Utility Airplane Council, a specialty group within the larger organization. The National Business Aircraft Association, formed in 1947, mirrored the early importance of business flying in the United States, and the light plane community began looking for a term to avoid the popular image of casual flying and project the respectable image of the NBAA. During the 1950s, general aviation became an accepted term, and the General Aviation Manufacturers Association became a separate entity in 1970. This institutionalization of the light plane industry marked a rite of passage.

The Postwar Airliner Market

Most of the major postwar transport variants from U.S. manufacturers were produced in numbers of 100 to 300; during the 1950s and early 1960s, the total number of aircraft operating on the world's scheduled airlines numbered about 3,500, with a total of 2,500 manufactured by companies in the United States. As the new generation of jets entered service during the 1960s, the total number of airliners increased some-

The Lockheed Constellation epitomized elegance in postwar international travel. Features like tricycle landing gear, pressurized passenger cabin, electronic navigation equipment, and other technologies were characteristic of the postwar generation of airliners. Also, increasing numbers of foreign airlines favored air transports built by American manufacturers. *Courtesy of the Lockheed Corporation.*

what, and American aircraft still represented the backbone of foreign lines as well as domestic airlines. Of some 4,000 transports in worldwide service in 1969, more than 3,000 (75.8 percent) originated in America.[1] While the United States could never be accused of achieving a total monopoly in the long-range airliner market, the dominance of American-made aircraft was certainly evident. Frustrated Europeans began to claim that the American position was the result of an unfair World War II agreement, in which European manufacturers concentrated on fighter production, leaving the States to build bombers and transports. This alleged agreement enabled Americans to shift quickly to production of large transports after the war—an unfair lead that Europe never seemed quite able to overcome.

Theodore Paul Wright, chief of the international bureau of America's War Production Board during World War II, discounted the European charge, noting that the patterns of production represented manufacturing strengths that already existed in the early 1940s. American production capacity and current designs made it logical for the Allies to rely primarily on transports and bombers from the United States, even though Britain continued to produce four-engined Lancaster bombers and other large aircraft during the war, actually converting many of them to long distance transports (the Avro York) in the early postwar years.[2] It should also be remembered that operators of the Douglas DC-3 had already captured the lion's share of passengers by 1939, leaving European airliners behind. Moreover, designs for the DC-4 began in 1935 and the Constellation in 1939; wartime production obviously helped, but superior U.S. airliners were already underway before the war began. More to the point, the self-contained American market alone far surpassed that of any single European nation, so that individual designs from Britain, France, or Italy in the postwar years could scarcely hope for the scale of production that would reduce unit costs against competitors and generate additional international sales. Not until the Airbus consortium of the late 1960s, with its promise of multinational sales to its own European partners, did a European airliner achieve the potential for the volume production advantages inherent in the North American marketplace.[3]

In any case, getting an airliner into service was no guarantee of commercial success. In the fall of 1945, the full impact of postwar cutbacks became evident to Boeing as production orders for the B-29 plummeted from 155 per month to 10, with prospects of further cuts. The company had commitments for a C-97 military freighter, a double-deck variation of the B-29, but nothing really tangible to shore up a payroll running at $500,000 per day. Company engineers thought the C-97 could be the springboard for a commercial passenger transport, although Douglas and Lockheed already held formidable positions in the airliner business. William Allen, Boeing's new president, took the plunge, putting engineers to work on an airliner version of the C-97. He pegged the price at $1 million each, and ordered a production run of 50 planes even though no orders existed for them. Then Allen sent a special message to his sales department, instructing them to get out and sell the B-377 Stratocruiser. The prototype flew in 1947 and entered service with Pan Am in 1948.

The Stratocruiser contributed unmatched luxury to air travel of the late 1940s and mid-1950s, and the Boeing sales force touted the plane's size and comfort as a means to compete with the blandishments of Pullman rail travel and the spacious decks of transoceanic passenger liners. Seating was unusually flexible, with accommodations for from 55 to more than 100, and some planes boasted 28 upper and lower berths for overnight transcontinental and intercontinental runs. Passengers especially remembered the 377 for its wide seats, legroom, and headroom. The bar/lounge, entered by a spiral staircase leading to the lower deck, was a sensation for early postwar airline travelers. For all its passenger appeal, the plane was expensive to build and Boeing sold them at a loss. The company realized a benefit, however, because it kept together an invaluable team of people in the technical and manufacturing force. Boeing sold 56 Stratocruisers before terminating the line in 1950 and did not earn a profit on them for 10 years, after a decade of selling spares and parts to airline operators. Boeing's search for success in the field of postwar airliners remained elusive until it made another gamble in the late 1950s, again extrapolating from its experience with big military aircraft.[4]

With more experience in airline sales and proven transport designs from wartime operations, Douglas and Lockheed seemed to fare more successfully. The Douglas DC-6 used the same basic wing of the DC-4, beefed up to take bigger and more powerful engines, and its pressurized fuselage included an extra seven feet of length. A subsequent variant of the DC-6 family, the DC-6B, acquired a reputation as one of the most efficient airliners ever built. Within a decade of its first flight in 1946, Douglas sold more than 600 DC-6 models that carried the colors of some four dozen airlines. Along with Lockheed's Constellation, they pioneered regular transatlantic schedules, although they still had to make regular refueling stops at Newfoundland or Ireland, especially on westbound flights against the prevailing winds.

On domestic routes in the United States, American Airlines president C.R. Smith wanted a plane to beat TWA's improved Constellations on prestigious coast-to-coast runs. If Douglas could produce a convincing proposal, Smith guaranteed enough orders to warrant the start of production. During 1951, a veteran design group at Douglas analyzed the challenge in view of the growing appeal of jet engines, then backed off, concerned about the jet's reliability and high fuel consumption for commercial airlines. Using Wright Aeronautical's latest radial engines, Douglas built the DC-7 as well as an advanced model, the DC-7C, used by Pan Am to offer the first true nonstop transatlantic routes.

Entering service during the late 1950s, the advanced piston-engined designs soon gave way to long-range jet transports on longer routes. Nonetheless, the postwar Douglas and Lockheed airliners transformed passenger flying in America and made intercontinental airline travel a new transportation phenomenon. In 1956, more travelers between America and Europe went by air than steamship; in 1957, domestic U.S. airline miles surpassed both trains and buses, making airlines the nation's leading intercity carrier.[5]

In the meantime, the airlines shopped for a modern, postwar airliner to replace the venerable DC-3 over shorter routes. Martin and

Convair both developed twin-engine transports, but the Convair 240/340/440 series, delivered from the start with pressurized cabins, enjoyed greater success. After a specialized modification firm in California, PacAero Engineering, developed a lively business of converting the Convair's robust airframes into twin-engine turboprops during the 1960s, Convair went into the business itself. Foreign manufacturers also began to exploit the obvious market niche for smaller transports. In 1955, the Dutch manufacturer Fokker developed the Friendship, an appealing, high-winged twin turboprop with a pressurized cabin. In 1956, Fairchild signed a production license and delivered 205 models of the F-27 (the U.S. designation) through the late 1960s.[6]

Other techniques sometimes helped recoup costs from what seemed to be a failed undertaking. With its C-130 Hercules military transport in production, Lockheed hoped to use it as the basis for a similar high-winged, four-engine airliner but failed to get a customer. Early in 1955, American Airlines solicited proposals for a larger turboprop, so Lockheed reworked its design as a low-wing airliner, the L-188 Electra. American and others ordered several dozen off the drawing boards. Lockheed leveraged these orders by turning out a third version of the airframe, an aerodynamic prototype for a Navy patrol plane that became the P-3 Orion. After the Electra transport went into service early in 1959, a series of nasty in-flight accidents led to severe speed restrictions that hampered its operations for two years. Investigators traced the cause to engine nacelle vibrations, and modifications restored the plane to successful service. But the restrictions obviously wounded the Electra's career, and the advent of smaller jets ended it after a production run of only 174 aircraft and an estimated loss of $57 million, plus $55 million more in damage suits. Hardly a positive performance on the balance sheets. The P-3 Orion, however, went into service in 1962 and stayed in production into the 1990s, with hundreds of Orion variants in service with many nations.[7]

The negative impact on the Electra's sales after its accident history underscored the need to continue selling air travel to the American public. Airlines maintained extensive advertising campaigns during the postwar years, with added support from aviation manufacturers themselves. Although manufacturers and suppliers had regularly paid for ads in trade magazines and specialized periodicals like *Aviation Week*, forays into the mass market signaled a notable departure in marketing strategy. The mass market venture paraded the manufacturer's name past the eyes of far more people, who might be encouraged to inquire about a particular make of airplane when booking reservations, and also served to popularize the whole idea of air travel, thus expanding the market for aviation manufacturers as well as the airlines.[8]

As a major travel periodical of the era, *Holiday* magazine was a logical choice for aviation-related advertising, and both airframe and power plant manufacturers bought space there. Eager to break into the postwar airline industry, the Glenn L. Martin Company became one such advertiser, and its advertisements are characteristic of the era. One full-page, color presentation in 1947 stressed airline travel as the way for busy Americans to save time, permitting longer vacations and more productive business trips. The ad listed airlines using Martin equipment and encouraged uncertain readers to write the Martin Company for a free book, "How to Travel by Air." Other manufacturers, including the Pratt & Whitney engine division of United Aircraft, were regular advertisers in general interest magazines, and Boeing remained an advertiser in popular magazines like *National Geographic* through the mid-1960s.[9]

The Jet Airliners

Beginning in the late 1950s, a second wave of change occurred when faster, long-range jets entered service. As in the case of gas turbines, the British again led the way in jet airliners when the de Havilland Comet I launched service from England to Africa and the Far East in 1952. Over the next two years, the Comets proved to be quiet, smooth-riding planes that enjoyed great popularity. Unfortunately, a series of tragic accidents grounded them for two years until investigators painstakingly traced the cause to metal fatigue induced by a seemingly inconsequential change in riveting procedures. The Comet's early success appealed to operators in the United States, and Pan Am had already placed orders for a larger version of the Comet, with deliveries due in the late 1950s. By the time the Comets were ready to start flying again, the American aviation industry had taken belated but giant strides to catch the British.

In 1952, the same year Pan Am placed an order for the Comet, a number of U.S. engineers decided the time had come to design an American jet airliner. Several factors converged to convince American companies that jet liners were feasible. As axial flow engines improved, their fuel consumption decreased to levels viable for airline operations. Industry leaders became especially interested in Pratt & Whitney's J-57, which took to the air in 1952 with the Boeing B-52 bombers. Also, engineers were coming to grips with problems associated with the design of large, subsonic planes, including aspects of low-speed control on swept-wing configurations. Boeing in particular had accrued considerable swept-wing expertise during the B-47 and B-52 programs. Military spending as a result of the Korean War yielded profits that Boeing could allocate to commercial jets. Besides, as the B-52 neared production, its design teams became available for an airliner project.[10]

Because Boeing's own wind tunnel was shut down for repair and improvement, Boeing leased test time in Purdue University's tunnel. This increased design costs, but the company forged ahead even though it had no firm customers on the horizon. Once design was complete, Boeing proceeded with testing despite the airlines' skepticism, and built a prototype using its own money. Boeing's accountants estimated the price of the jet transport prototype at $15 million or more—roughly four times the company's profits since the end of World War II. The initial design authorization used the designation 367–80, and the big new jet soon became dubbed the "Dash 80." Boeing's extensive experience with the B-47 and B-52 programs enabled the Dash 80 project to progress quickly, with the first flight on July 15, 1954, two years after Boeing's board of directors had given its approval. At the same time, Boeing saw a way to hedge its risks by launching parallel development of jet-powered tankers for the B-52. In presentations to the Air Force, Boeing argued that only a jet tanker promised the range and speed necessary to manage refueling for fast, high-flying planes the size of the jet-propelled B-52 bombers. Eventually, the Air Force agreed and awarded a contract in 1955 for the development and production of the KC-135 jet tanker.[11]

Boeing still needed an airline customer. Most of the airlines still looked skittishly at jets; they had already placed extensive orders for advanced versions of Constellations and the Douglas DC-7C, capable of crossing the Atlantic nonstop in any kind of weather. As a next step, most airlines planned to use turboprops. Juan Trippe of Pan Am was the exception. Trippe enjoyed playing manufacturers against each other in order to extract the best airliner at the best price for Pan Am. There followed a truly byzantine series of negotiations as Trippe wrested an agreement from Douglas to build a bigger jet transport, the DC-8, to challenge Boeing and the Dash 80 (now called the 707). To power the Douglas jet, he got Pratt & Whitney to agree on production of a powerful new engine, the J-75, after hinting at a deal with Rolls-Royce in Britain. Alarmed at the prospect of losing the order, and with haunting memories of the Model 247 failure, Boeing made a new pitch to Trippe, giving him the opportunity to use the 707–120 for the first jet service across the North Atlantic (with refueling stops), followed by quick deliveries of subsequent planes on order for Pan Am. At favorable prices to Pan Am, Boeing also promised early completion of a long-range Model 707–320, built with a 15-foot longer wingspan, a different wing planform, and a fuselage stretched by eight feet that could accommodate 24 more passengers than the 120. All this meant that Boeing had to rework its basic 707 fuselage configuration to a size larger than its tanker and maintain separate production lines. The costs for jigs and other equipment increased, creating a threateningly large deficit for several years, although the effort resulted in a more successful family of commercial and military planes that eventually outsold its Douglas rival.[12]

The 707 went into service on the New York–London route in 1958. Boeing thus got its jet airliner into operation ahead of its competitor and captured the lion's share of headlines as well as of the market. The DC-8 became a popular and successful design, too, with more than 550 planes delivered (Boeing 707 production eventually reached the 800 mark). During the early 1960s, Convair produced a similar design, the 880/990 series, but deliveries fell far short of the Douglas and Boeing rates, forcing Convair to cease its manufacture after completing 102 planes. Meanwhile, de Havilland's redesigned Comet IV went into service with BOAC in 1958 and, in a race for prestige, made the first jet passenger flights across the North Atlantic three weeks ahead of Pan Am. But the American jets boasted better range and speed and soon dominated jet sales to airlines around the world. The big new jets demanded a high-stakes gambling mentality: Douglas wrote off $298 million in development costs and production losses; Boeing expenses came to $165 million; Convair lost $425 million.[13]

European manufacturers found it frustrating to compete with the Americans despite the advantages of smaller staffs, lower wages, and comparably less overhead. During the 1950s, their products seldom competed effectively against American rivals. There were some exceptions, especially in cases where the European products filled a particular market niche or appeared on the market first. Examples included the four-engine British Viscount turboprop, the twin-turboprop Fokker Friendship (F-27) from the Netherlands, and the French Caravelle twin jet. The Caravelle made its first flight in 1955, entered European service in 1959, and appeared on routes in the United States during 1961, when United Air Lines began flying it. Chronologically, the Caravelle and other early British jet transports established the milestones for American competitors to follow in building jet airliners. The eventual success of American designs in the marketplace can be attributed to a combination of factors: bigger production runs and lower unit costs, aggressive sales campaigns, close attention to service and repair, and generally more powerful engines that enhanced performance and made it easier to stretch basic designs to accommodate more passengers, a factor that contributed to all-important revenues per plane.[14]

Experience with the big jets made airline operators more confident about adding smaller jet equipment on shorter routes to smaller cities. Designed for such shorter routes, the French-built Caravelle pioneered a design using a "clean wing," with an engine mounted on either side of the tail rather than using conventional wing mounts. This allowed engineers to build a wing for maximum lift without having to make allowances for engines installed on the wing itself. After exhaustive design studies, Boeing considered engines in the tail of its new transport, the 727. Responding to suggestions from TWA for a three-engined plane with all of the engines mounted in the tail, and concerned about strong European orders for the Hawker Siddeley Trident, designed precisely to that specification, Boeing formally announced its own three-engined Model 727 in 1960.

The 727 was carefully planned to operate on short-to-medium stage lengths and for service to and from smaller airports. It had important self-contained features like boarding stairs and ground support systems. Pushed by Eastern, Boeing designers dropped early plans to use a Rolls-Royce engine and adopted the Pratt & Whitney JT9D, preferred by Eastern for its rugged design and potential for increased power. But the 727's strongest appeal lay in Boeing's high-lift wing system, which utilized an arresting array of leading-edge and trailing-edge slots and flaps for enhanced lift at the lower speeds required by smaller airfields. The plane's flexibility for use on a variety of routes and airfields made it popular across a wide market.[15]

Boeing was able to keep some costs under control through its frugal habit of "commonality," using the upper lobe of the Model 707 fuselage to save tooling costs on the 727. Nonetheless, original estimates of a price of $3.5 million per plane escalated to $4.2 million when the plane entered service in 1964. The costs of the new postwar jets kept climbing, partly because of larger size, new engines, and modern alloys. But much of the cost probably stemmed from habits the manufacturers acquired while working on defense contracts, especially the practice of assigning extra-large engineering teams to military programs characterized by complex weapon systems. These expensive habits carried over to the new passenger jets such as the Boeing 727, whose development costs were so high that Boeing had to sell at least 300 planes just to break even—a sales total that represented the entire production run of early models of the DC-6 and Constellation. Fortunately, the thorough testing and painstaking engineering on the 727 paid off for Boeing, which had invested $150 million in the project, as well as for the five initial customers that had committed $700 million in orders and options before the plane's first flight. The plane performed better than predicted; fuel mileage, for instance, was 10 percent better than anticipated. Boeing delivered its last Model 727 in 1984, marking a total of 1,831 units—at the time, an unprecedented production record for postwar airliners. Considered a huge gamble at its inception, the 727 became a cash cow for Boeing for some 20 years.[16]

Douglas also decided to build a short-haul airliner, choosing a design closer to that of the Caravelle. The DC-9 made its first flight in 1965 and entered service the same year. Like Boeing, Douglas wanted to offer a self-contained airliner well suited for service to smaller cities that lacked convenient ground facilities. Even though the DC-9 was a smaller jet, its research and development costs were considerable. Faced with this issue, Douglas led the industry in developing shared-risk production, in which a major partner (or partners) agreed to carry tooling and production costs for primary parts of the airplane, sharing the risk and potential profit with the prime contractor. For the DC-9, de Havilland of Canada agreed to take responsibility for production of the wings. Early sales went slowly, with Douglas and its colleagues losing as much as $500,000 on each delivery of the $3.3 million airliner. The DC-9 recovered and became a success, with more than 700 sold by the mid-1970s and an additional 700 on order. Varying models offered customers a choice of planes accommodating from 90 to 139 passengers, and exhibiting different performance characteristics for short-field operations. Like the 727, the DC-9 became an important source of continuing revenue for Douglas.[17]

The postwar jets brought additional changes to the character of air travel, domestically as well as globally. The Douglas and Boeing planes, in particular, opened up travel opportunities to formerly remote corners of the world. The DC-8 and 707 transports carried from 150 to 180 passengers on long global flights, and, as British journalist Clive Irving put it, "On the whole, they did it dependably, safely, and with relentless frequency. This was an American-led revolution. And, being American, it had little patience with elitism." Elitists might bemoan the passing of the measured elegance of travel aboard steamships, but the jets helped expand global travel with an egalitarian style that was revolutionary. Essentially, the revolution was based on the Boeing swept-wing layout for the B-47 and B-52, translated into the quintessential form of the 707 and progressively polished over the following decades.[18] American manufacturers then recast this experience with large jet airliners and applied it to an equally successful line of medium-sized jet airliners like the DC-9, 727, and others.

The General Aviation Sector

In the postwar years the conventional definition of "general aviation" was clear enough, but when traced down through succeeding layers of federal nomenclature and weight classification, it became increasingly complex. In the simplest terms, it is necessary to keep in mind that the term "general aviation" covers all civil aircraft and aviation activity except that of certified air carriers. References to the "general aviation industry" in terms of the kinds of planes produced by manufacturers can be confusing. Technically, the term "light plane industry" would be preferable, since large transports operated by the airlines are considered as "transport category aircraft" if they have a maximum takeoff weight of 12,500 pounds or more. The latter type was built by Boeing, Douglas, Lockheed and others, leaving everything else to light plane builders such as Beech, Cessna, Piper, and many more. However, as subcontractors, the light plane companies often turned out components for heavy military planes as well as large transports. Moreover, if a surplus military plane or a passenger airliner was converted for use as a fire bomber in forestry work or as an executive transport for a corporation, they entered the general aviation fleet. At the same time, some twin-engine light planes entered service on scheduled routes as commuter airline transports; since they weighed less than 12,500 pounds, they were not bound by Civil Aeronautics Board regulations regarding routes and fares. Thus, the commuter lines also operated as part of the general aviation fleet. In the text that follows, the light plane industry refers to products under the 12,500-pound limit; the terms general aviation fleet, industry, and sector refer to all of the aircraft and economic factors that attend their operations.

The airliners and general aviation aircraft shared some 13,000 airports by the early 1970s, although the scheduled airlines themselves served only the larger airports of approximately 400 cities. Of these, about 25 major cities generated 70 percent of all airline traffic, and 150 large urban locations accounted for some 90 percent of the airlines' business. At the same time, industry and commerce in the United States exhibited a growing trend towards decentralization to smaller urban areas. Just after World War II, more than half of all U.S. manufacturing plants operated in cities of 100,000 people or more. As decentralization gathered momentum—driven by market considerations, taxes, available labor, and other factors—more companies either relocated or established subsidiaries in smaller cities and towns. By 1956, one-third of all new factories went up in smaller communities, usually where scheduled air service was nonexistent or infrequent at best. These were the localities represented by more than 12,000 airfields that were accessible to personal and corporate airplanes.

Decentralization constituted an important element in the rise of general aviation activity. Annual production of the light plane industry ran at about 3,000 units through the mid-1950s, rising to 7,000 per year by 1962, nearly 16,000 in 1966, and trailing off to 12,500 units delivered in 1969. From the unusually high figure of $110 million in net billings during the boom year of 1946, annual figures plunged to less than $20 million through the rest of the 1940s, recovered sharply to $100 million by the mid-1950s, and reached a high of $639 million in 1969. Exports claimed a substantial share of annual billings, ranging from 15 percent in the 1950s to 20 percent in the late 1960s, when 2,000 to 3,000 aircraft per year went to foreign customers. Both American and foreign aircraft represented a large market for light plane engines, an area virtually monopolized by Continental and Lycoming in the United States, where the unusually broad market gave them the production capacity and low unit costs to dominate this field. Airframes were far less difficult to build and required lower tooling costs; potential entrants in the engine market faced formidable hurdles.[19]

The general aviation fleet in the United States rapidly proliferated after the war, despite occasional downturns in production. Based on FAA records, the fleet numbered 37,000 in 1945, reached 58,000 in 1955, climbed to 95,000 10 years later, and stood at 130,000 in 1969. The magnitude of the general aviation total stood out in comparison to some 2,400 active U.S. commercial airliners on scheduled routes and about 30,000 military planes at the end of the 1960s. Although the total dollar value of general aviation's annual production fell far short of the totals for airliners and military aircraft, the general aviation market played an important role nonetheless. In their thousands, general aviation planes conducted myriad operations in flight instruction, crop treatment, surveying, photography, ambulance duty, and other tasks. The lion's share of passenger miles were racked up by business people carried in single-engine personal planes, twin-engine executive planes, and corporate jets. Because American-built light planes made up most of the worldwide general aviation fleet during the early postwar decades, many individuals in the aviation business in other countries developed an early experience with United States products; this often served as a benefit in winning multimillion-dollar contracts for foreign sales of American airliners and other aviation equipment.[20]

As early as 1943, when all aircraft companies were backlogged with military contracts and D-Day was still in the planning stages, manufacturers began looking ahead to a postwar bonanza in private flying. Many aviation journalists joined the bandwagon, including John Andrews, an editor for *Air News* magazine, which purportedly kept him on top of aviation manufacture and design. His book, *Your Personal Plane*, was published by the respected firm of Duell, Sloan and Pearce in 1945. With an introduction by former Vice President and then Secretary of Commerce Henry A. Wallace, the book offered a highly optimistic prognosis for postwar aviation, predicting 350,000 planes to be built over a four-year period. Among other things, Andrews argued that a $1,000 plane was affordable for anyone who could spend $40 per month in annual expenses, about the same for "the average light six-cylinder automobile." Later, the author revealed that the $1,000 plane would accommodate only two people; still, the retail price of $1 per pound was only a little higher than the ratio in the auto industry. Keeping that fact in mind, most families would find it prudent to spend $3,000 for a comfortable four-place plane, a product that would "give the young parents of America a flying family car for less than they will pay for a Buick Sedan."[21]

During 1945, light plane advertisers began to expand their appeals outside the aviation trade journals. Piper placed ads in *Life* magazine for its Piper Cub ("Points the Way to Wings for *All* Americans") that featured sequential cartoon drawings illustrating how easy it was to take off and land from any convenient airstrip, even a rough or soft field. The ads also offered booklets and films on "How to Fly," as well as a free pamphlet, "What Your Town Needs for the Coming Air Age." Other light plane manufacturers ran ads in mass circulation periodicals ranging from *Business Week* to *Better Homes and Gardens*. A notably upbeat article in the February 1946 issue of *Fortune* magazine ("New Planes for Personal Flying") gave an overview of proposed models and planes in production, enthusiastically proclaiming that "this year the buyer of personal airplanes has the widest choice in aviation history." The prices in *Fortune's* catalog ranged from $1,885 for a Piper Cub to $60,000 for a Twin Beech. The article admitted that though prices were still an obstacle, costs were on the way down.

But costs went up. In 1946, everybody in the business set production records when the light plane industry sold 33,254 aircraft, 455 percent over the last prewar annual sales figure. Industry observers were understandably encouraged. With increased demand, higher volume, and mass production, prices would come down, and even more planes would be sold to all those veterans expected to enroll in flying schools.[22] The euphoria about the possibilities for general aviation activities led some of the big wartime manufacturers to enter the field of light plane production. On the whole, they fared badly, soon abandoning this market to the traditional companies who specialized in such aircraft. A large number of smaller manufacturers also tried to find a niche in the postwar general aviation market, and on the whole they failed as well, with two or three exceptions.

In 1943, Republic Aviation, builders of the rugged P-47 Thunderbolt, bought the design and flying prototype of an unusual amphibian they called the "Seabee." Promotional brochures depicted contented pilots anchored in some remote, idyllic cove, with the pilot tending a fishing pole while lounging in one of the Seabee's front seats. Production models began deliveries in early 1946, priced at $3,995, and thousands of orders rolled in. As the company became aware of escalating production costs, it apologetically raised the Seabee's price to $5,995. During 1947, a thorough accounting analysis revealed actual expenses of $13,000 per plane on deliveries of just over 1,000 units. Republic killed the Seabee and returned to military production.

On the West Coast, North American Aviation went through a similar experience. The company's four-seat Navion evolved from a concept developed by the original P-51 Mustang design group. Research and development recovery costs ran at twice the 1947 price of $7,750, so North American sold the Navion to Ryan Aeronautical, who made slim profits on a proven design until the Korean War ended production in 1951. The inability of large manufacturers to produce small planes at a profit underscored the particular challenges of the general aviation market.[23]

Still, determined entrepreneurs kept trying for aeronautical success, caught up in the postwar optimism that predicted a plane in every garage. In this regard, several inventors built and flew roadable airplanes. Characteristically, these vehicular compromises featured detachable propeller, wings, and rear fuselage stored in the family garage, attached for a flying trip, and left behind at the airport while the owner blithely drove off in the cockpit/passenger module. The trade-offs for practical operation on both highways and airways proved too contradictory, and none of these efforts achieved commercial success.

Of the many traditional light plane builders active before the war, few survived. Inadequate financing, inefficient manufacturing, outdated designs, poor marketing—a variety of factors took their toll. Waco, a 1930s leader, suddenly collapsed. Stinson was absorbed by Piper. Luscombe persevered through the 1950s under different ownership in different states, but eventually folded. The most successful of the survivors—Piper, Cessna, and Beech—stayed alive through ongoing modernization of their products, and diversification into upscale twin-engine planes for assorted business and corporate operators. Their existing network of dealers played a significant role in their survival, offering a base of customer loyalty and continuing product support for owners who traded up to more expensive planes.[24]

Meanwhile, the domestic market went sour. Sales fell off in 1947 to 15,617 units, skidded to 3,545 in 1949, and plummeted to their nadir in 1951, when only 2,477 planes were sold. The causes were varied. The less expensive two-place planes, with a 95- to 100-MPH cruising speed and docile flying characteristics, were less than exciting for many ex-military fliers. Besides, there were numerous other options coming on the market, in the form of 31,000 war surplus aircraft. Included among these were many Twin Beeches, C-47s, and twin-engine bombers that had been modified into executive transports. Novice civilian student pilots comparing the style and comfort of Detroit's postwar cars found many postwar light airplanes noisy, drafty, and cramped. Light planes were easily grounded in bad weather, and operating costs were discouraging for the majority of families planning to fly on weekend junkets. Moreover, few resort and vacation centers were close to convenient airstrips. In rapid succession, factors such as Cold War tensions, the Berlin crisis, and the outbreak of the Korean conflict meant further damping of civilian light plane sales.

With more realistic notions about the nature of the market, refinement and improvement of aircraft, and development of flying facilities, the light plane industry managed a slow but stable recovery. Manufacturers began to focus more attention on designs for the business pilot, producing aircraft with greater comfort, reliability, speed, and range. Aircraft that met speed and range requirements, in particular, became available in several cost and performance categories, including a new generation of twin-engine planes.[25] Some of them were under development even as the downswing in the industry was gathering momentum.

The postwar experiences of Piper were typical of many manufacturers. The company completed its 1945 fiscal year in September with $7.7 million in sales. In the space of a few months, the backlog of orders climbed to more than $11 million, dealers had to be rationed, and the three-place Piper Cruiser was being sold on the black market at several hundred dollars above the factory list price. Managers at Piper confidently increased production. When the light plane market began its violent contractions in 1947, Piper discovered that the industry had overproduced. Many buyers had ordered from two or three companies, then taken delivery on the first plane available, canceling the other orders. But Piper was also guilty of continuing to market designs that had changed little in one and a half decades. The firm's small, fabric-covered planes were suitable for sunny-day flights in visual flight rules (VFR) conditions, but had limited equipment for night flying or operating in reduced visibility. Piper nearly went under. The company survived through draconian economic measures and a drastic reshuffling of its management. In time, the company found its niche with improved versions of the dependable Cub serving as trainers and versatile utility aircraft. The four-place Pacer, with a 115-hp engine, introduced in 1949, became the Tri-Pacer (tricycle gear) in 1951; the series also won praise for well-equipped interiors and instruments to enhance business flying.[26]

Cessna competed with Piper in the training and sport-plane categories with the 120/140 series, an 85-hp aircraft in deluxe or standard versions. More important was the all-metal fuselage of the 120/140 series, a harbinger of the future for airplanes of this class. When the postwar sales slump hit Cessna, company managers commissioned a series of market surveys that definitely pointed to a strong market in four-place business aircraft.

Introduced in 1948, the 170/172 series evolved into an all-metal design with tricycle landing gear, and it became one of the best-selling light plane series of the era. Cessna never underwent the wrenching managerial reorganization and financial problems of Piper, although the company occasionally scrambled for liquidity. Cessna found part of the answer in diversification. In 1947, the company took on an Air Force contract to build wood and aluminum furniture, a venture that took up part of the slack in sales for three years. Cessna also branched out into industrial hydraulics, a profitable operation that was retained.[27]

Beechcraft, like everyone else, entered the postwar era with gusto. The company realized that the biplane configuration of the Staggerwing, with its wood and fabric construction, symbolized a prewar hallmark that seemed out of place in the new era of metal monoplanes, and the Staggerwing production line finally closed down in 1948. In the interim, Beechcraft, too, felt the sales drought but weathered it better than its competitors. Demand for improved versions of the Twin Beech held fairly steady, and the company's traditionally conservative fiscal habits protected the surplus carried over from wartime manufacturing contracts. Beechcraft's products cost a bit more, but the company catered to a narrower market than did its competitors, and Beechcraft owners were aggressively loyal. Most important was Beech's early postwar development of a successor to the single-engine Staggerwing. The manufacturer had a reputation for producing high-performance aircraft and shrewdly planned to continue this tradition with a new ultramodern design, while keeping the price as reasonable as possible. The new plane was called the Model 35 Bonanza, an apt sobriquet for the sales it eventually generated.

The Bonanza featured a number of advanced design and construction features. Sales presentations also stressed the plane's generously sized four-place cabin, extrapolated from family sedans of the era and carefully soundproofed. Interior design and the instrument panel reflected the latest theories of industrial psychology. Bonanzas came with radio gear and modern instruments for serious instrument flying. Finally, there was a distinctive and unique V-shaped tail, a feature that gained wide acceptance as a Beechcraft postwar hallmark. Offering a 175-MPH cruise speed and 750-mile range, the Bonanza introduced a new era in single-engine light planes and in business flying.[28]

The early postwar era also witnessed the introduction of a new generation of twin-engine aircraft, some with speeds of more than 200 MPH. At the end of the 1950s, twins could generally be grouped into three categories—light, medium, and heavy (high performance)—based on their passenger load, range, horsepower, and speed. Prospective buyers might look, respectively, at the Piper Apache, Aero Commander, or Twin Beech. The same categories characterized single-engine designs as well, reflected in types like the Piper Tri-Pacer, the Cessna 170, and the Bonanza, respectively. Over the years, each category encompassed numerous additional types, offering buyers a choice of planes for specifically defined operational requirements and operating environments.

With this new range of postwar aircraft, the light plane industry began to make its mark. Many skeptics still regarded the industry's products as marginally safe—little planes with questionable reliability over short, inconsequential hops. Convenient cross-country flying on instruments became a reality after "very high frequency omni-range," or VOR, units were installed in the early 1950s. Although the VOR system had been planned for scheduled airline use, electronics companies like Narco soon developed low-cost, lightweight equipment for the light plane industry. Private pilots began using the VOR network, giving general aviation aircraft extraordinary new versatility for night flying and operations in messy weather. Improved, static-free ultrahigh frequency (UHF) radios also came on the market, and Bill Lear introduced a light plane autopilot system that eased the strain on pilots whose flights were becoming longer and more frequent. The term *general aviation* in the postwar decade replaced *private flying* in a move to get away from the aeronautical country-club image that many felt the latter term implied. In the 1950s, when President Eisenhower's personal pilot began using an Aero Commander to fly the nation's chief executive between Washington and his farm near Gettysburg, Pennsylvania, it seemed to epitomize the maturity and legitimacy of the general aviation sector.[29]

By common understanding in the growing aviation industry, general aviation included all aircraft and operations not related to military flying or the scheduled commercial airlines—a bewildering variety of aircraft and flying activities. The airplanes included not only those manufactured by the light plane industry, but gliders, home-built planes, and considerable numbers of ex-military planes converted for civil use. These ran the gamut from biplane Stearman trainers to twin-engine Douglas B-26 Invaders and Boeing B-17s, which were reworked by several specialized modification companies. The Stearman and other war surplus biplanes were modified for crop dusting and seeding. Beginning in the early 1950s, several dozen Invaders were modified by On Mark Engineering, outfitted with passenger seats and customized interiors to be flown as high-speed executive transports. Other postwar aircraft were converted into "fire-bombers" for dumping fire retardants on forest fires, carrying parachuting "smoke jumpers" to forest fires in remote areas, or operating as spray planes. A number of specialized businesses also car-

Cessna's Skyhawk typified successful postwar general aviation designs for pleasure and business trips. As floatplanes, such aircraft multiplied their usefulness for carrying cargo and passengers to otherwise inaccessible areas. Floats were often fabricated by such longtime suppliers as the Edo Corporation. *Courtesy of Cessna Aircraft.*

ried out modifications of standard light plane designs to achieve better speeds, short-field performance, and operational efficiency.[30]

In terms of production volume, Cessna emerged as the leader in 1955, when the company not only outsold its rivals but delivered more new airplanes (1,746 commercial and 24 military) than any other company in the world—a distinction that the company maintained year after year through the 1960s. Cessna produced a dependable product, but also spent more time on marketing techniques, wooing customers with ads that aped the gushy exuberance of the auto industry. In the pages of *Flying* magazine, the Cessna 180 became "1954's Smoothest Airplane," with "Living Room Comfort Aloft" in the cabin and an engine mounted and tuned so well that a full water goblet on the cowling didn't spill a drop while the engine was running. New color-striping, new upholstery, new sloping map compartment, new exterior baggage door—and, not just any old pair of wing flaps, but Cessna's own "Para-Lift Flaps"—were glowingly extolled. All for "only $12,690!" Two years later, it was "Land-O-Matic" landing gear for the Cessna 182—"the big new feature that makes flying like driving!" (Essentially, this feature involved a tricycle landing gear.) Later 182 models boasted of refinements such as "polycloud" seat cushions.[31]

Like Cessna, Piper refined its line of light singles, added some new twin-engine aircraft, and developed a new series of high-performance, single-engine planes, including models with retractable landing gear. In this respect, the company began to compete with the Beechcraft Bonanza, although the most talked-about new plane in this category came from Mooney Aircraft. After many years of activity in Kansas, Mooney relocated to Kerrville, Texas, in search of lower overhead and lower taxes. The company's simple, single-seat "Mite" evolved into an all-metal, four-place series. Aerodynamically refined around a snug cabin, the Mooneys in the 1960s annually accounted for more sales than any other plane in their class. Al Mooney and his brother had retired from the company in 1955, and the company subsequently changed direction several times. But as long as Mooney Aircraft concentrated on one or two versions of its basic single-engine model, the organization survived as a corporate entity into the 1990s.[32]

An especially notable trend within the industry involved the continued development of "heavy twins," incorporating advanced equipment like pressurized cabins and turboprop engines. The postwar development of light twins, while increasing the pilot and passenger accommodations for up to six or eight seats, still put everyone in cabins of comparatively limited size. In many ways, the Grumman Gulfstream I, priced at $1 million, represented the ultimate of the new breed of heavy twin, designed and marketed with the top-level corporate executive in mind. Skeptics predicted sales of only 15 to 17 planes, arguing that too many examples of the DC-3, costing $200,000, were available. Deliveries began in the spring of 1959; 60 planes were in corporate or personal service by the end of 1960, and more than 180 had been sold by the spring of 1966. Designed for up to 24 seats, the executive version of the Gulfstream I carried 10 to 12 passengers and a crew of two in luxurious comfort at speeds of 350 MPH at over 30,000 feet—the first of the new heavy twins to combine turboprop engines and a pressurized cabin. Its opulent interiors and customized appointments set the standards for executive flying. As in the case of Mooney, the fact that the general aviation market had attracted a successful new competitor was in itself a significant event. Beechcraft, which had dominated this end of the market for so long with the Model 18, responded with a new series of cabin-class twins known as the Queen Air and King Air, including pressurized, turboprop models.[33]

Mounting sales of aircraft in the heavy twin class demonstrated that corporations were willing to pay from several hundred thousand dollars to more than a million dollars for high-speed executive flying. The obvious question was whether there was a similar market for jets. Jets were expected to be much more expensive to buy and to operate, thereby limiting sales. Potential manufacturers were understandably leery because they anticipated high R&D costs and worried that a presumably limited market might result in considerable losses—especially if competitors got into a scramble for buyers. In 1961, two aerospace giants, Lockheed and North American, each introduced small jet transports for possible military training and for use as corporate aircraft. With price tags between $1 million and $1.5 million, sales fell far short of the break-even mark for either, and skeptics argued that a viable market for executive jets simply did not exist.[34] Others felt the market had not really been tapped, and that the answer was a less expensive airplane. One of the most insistent of these jet enthusiasts was a maverick named William P. Lear.

Before World War II, Lear marketed a successful car radio and other compact electrical products; wartime contracts involved radio and navigational equipment. After World War II, he developed an innovative company that pioneered a number of electronic devices, including stereo tape systems for cars. In 1949, he received the prestigious Collier Trophy for having designed the first successful autopilot. During the mid-1950s, Lear made a reputation in the corporate aircraft business by converting Lockheed Lodestars into long-range executive transports called Learstars. By 1959, he was already thinking about a radical new executive jet that would be fast, efficient, and small. Controlling the size would keep the price down, and making it speedy would appeal to corporations who saw it as a time-saving business machine, not a luxury liner.

The configuration was laid out by a Swiss engineering team using the existing wing design from a canceled Swiss military plane. Because Switzerland was centrally located in Europe and possessed excellent cargo connections, Lear considered it an excellent location to assemble components that were inexpensively manufactured elsewhere. But work went slowly, and the tangle of international suppliers kept schedules in a turmoil. Lear abruptly moved the whole project to Wichita, Kansas, in 1962. The move not only put Lear closer to primary American markets but also offered the production skills and technical base needed for the Learjet, as the plane was called. In Wichita, Cessna's prior work on the T-37 jet trainer helped to provide a nucleus of suppliers, production managers, and factory workers with experience in small jet aircraft. The city arranged attractive tax benefits for Lear's company and had low labor costs. A new plant went up in remarkably short order, and the first Learjet streaked off the runway in the autumn of 1963.

William Lear pioneered the development of efficient corporate jets. The Learjet 35A, shown in the background, recalls the earlier compact models introduced in the 1960s. The Learjet 55 in the foreground was advertised as a "wide-body" design, with various models introduced in the 1970s and 1980s. *Courtesy of Gates Learjet.*

Lear's fleet, executive jet carried two pilots and six passengers at more than 500 MPH to altitudes of 45,000 feet, statistics considerably above those of its competitors. Moreover, its price tag of about $600,000 to $750,000 amounted to approximately half that of the Lockheed Jetstar, and two-thirds that of the North American Sabreliner. By the autumn of 1964, only a year after the Learjet's first flight, the first production aircraft went to a corporate owner, with additional deliveries and sales passing the 100 mark in just 12 months. A savvy sales strategy helped make the sleek Learjet synonymous with air travel for modern sophisticates. Lear's plane appeared in the background of numerous ads featuring stylish goods and services, and also provided transportation for the high-style operations of the private investigator in *Our Man Flint*, played in the motion picture by James Coburn. Sales success, however, created problems in maintaining production rates and dissension among dealers anxious to finalize sales contracts. In 1967, Gates Rubber Company bought control of Lear's business, eventually known as Gates Learjet. Bill

Lear himself continued to work on electronic systems and other aviation projects. Meanwhile, sales of Learjets picked up after its reorganization, and competitors from America and abroad soon entered the burgeoning market. The era of executive jets had truly arrived, and, as with the advanced heavy twin, it was thanks to a company outside the circle of traditional general aviation manufacturers.[35]

At the other end of the performance (and glamour) spectrum, the design of airplanes for agricultural operations entered a new era based on research and development at Texas A&M University. Few projects had as much impact on the subsequent design of one type of aircraft as a product from Texas A&M called the Ag-1. Dozens of war surplus biplanes, like the Stearman, had been converted for crop dusting and spraying. They were rugged old birds, but not intended for low-level maneuvers while heavily laden with chemicals. Many planes stalled, and many pilots died. The project at Texas A&M got started in the early 1950s when George Haddaway, publisher of an aviation magazine in Texas, persuaded several light plane manufacturers to support design and construction of a custom-built agricultural plane.

The Ag-1 prototype took full advantage of postwar improvements in aerodynamics and fabrication. Special safety features included location of the cockpit for maximum visibility and protection in case of a crash. Out of this collaborative project, Piper and Cessna evolved designs of their own. The new planes reduced the accident rate by half and cut the fatality from one in every 20,000 hours of flight to one in every 90,000 hours. Piper's Pawnee was soon at work in more than 90 countries and the original Ag-1 configuration became the model for ag planes around the world. One American entrepreneur, Leland Snow, had helped design and build the Ag-1 prototype as an engineering student at Texas A&M. With his own savings as an ag pilot, plus deposits from other interested operators, Snow designed and built the S-2. In 1965, Snow completed his 300th airplane; 100 of them had gone to customers in Latin America, Africa, New Zealand, and other regions abroad. Their bigger payloads and operational radius not only kept them competitive with major producers like Piper and Cessna, but also kept them in production long after Piper and Cessna had closed manufacture of their versions.[36]

Helicopters represented another of the fascinating aviation technologies of the postwar era. In civil applications, they herded cattle, sprayed crops, carried executives, performed emergency rescue and medical service, served as broadcasting posts for radio traffic reports, scouted for ocean fish, transported news reporters, patrolled high tension lines, and more. Other increasingly sophisticated helicopters eventually complemented the original Bell Model 47 for civil use, although

The Bell Jet Ranger and other helicopters became widely used in the postwar era for an astonishing variety of tasks, including aerial ambulance missions, fishery spotting, cattle herding, construction, forest fire patrol, crew transfers for offshore rigs, law enforcement, and on-the-spot news coverage. *Courtesy of Bell Helicopter.*

they never completely replaced it. The next major civil type produced by Bell was the Model 205, a turbine-powered (turboshaft) design with 14 seats based on the military UH-1 Iroquois of 1961. The Model 205 silhouette became a basic pattern for later and more successful helicopters like the Model 206 Jet Ranger. Much smaller than the 205, Jet Rangers seated the pilot and four passengers and proved to be economical and practical aircraft, serving in a variety of executive, commercial, and utility versions. Built in America and overseas, they undeniably helped establish a growing market for civilian helicopters from the mid-1960s onward.

Production of various types of the Model 47 and its successors amounted to several thousand by the end of the 1960s, including dozens of civil units produced under license by Agusta in Italy. Hiller and Hughes also delivered several hundred more light civilian helicopters to

American and foreign customers. Encouraged by this activity in the civil sector, and extrapolating designs from their successful military production lines, Sikorsky and Boeing Vertol each developed a large passenger helicopter intended for commercial transport duties as a 25- to 30-seat shuttle between urban centers and major airport terminals. High operating costs and flight restrictions in heavily populated urban areas proved discouraging. An amphibian version of Sikorsky's helicopter, S-61N, eventually sold more than 100 versions when the company's marketing organization exploited a worldwide demand in the petroleum industry to ferry cargo and personnel to offshore oil rigs and remote construction sites. Helicopters for civil operations became commonplace, and small-to-medium designs continued to build a moderate but steady rate of sales.[37]

Featuring pressurized passenger cabins, stretched fuselages, bigger engines, and other improvements, the piston engine airliners of the early postwar era made intercity and international air travel increasingly popular. With broader marketing and rising global demand, manufacturers of airliners enjoyed this welcome source of revenue. After the military services adopted jet aircraft, the airline industry cautiously moved into the gas turbine era. A variety of American-made jet airliners, boasting unusually compelling lower unit costs because of large orders from the North American market, eventually dominated the world market, as well. Strong sales of jet equipment continued because of the jet airliner's low operating cost and high utilization rate.

Often overlooked, the general aviation industry flourished after a rocky start in the late 1940s. The continuing decentralization of American business during the postwar decades made it necessary for business people to find rapid, convenient travel to thousands of localities outside the market areas of scheduled airlines. The general aviation industry willingly supplied such transportation, with manufacturers turning out a surprising variety of designs, including helicopters. Offering a broad array of aircraft types that often benefited from advanced aerodynamic and fabrication techniques of the era, U.S. manufacturers dominated the world market for general aviation planes and engines. Pressurized corporate transports and executive jets became commonplace elements of business operations in America and overseas, swelling the market for general aviation aircraft, engines, and accessories. For military hardware, airliners, and general aviation alike, the sales and manufacturing ventures became more and more an international enterprise.

CHAPTER 7
Changing Horizons 1970s/1980s

MOVING INTO THE 1970s and 1980s, a thaw in Cold War antagonisms brought a major shift in the deployment of thousands of nuclear-tipped missiles as the United States and the Soviet Union agreed to reduce numbers. This shift in the field of military rocketry was accompanied by changes in America's space exploration programs, principally in terms of growing trends toward international collaboration. While such international efforts were not new, budgetary constraints in the American space program and a growing capability in European aerospace technology engendered far more cooperation in manned as well as unmanned missions.

During the late 1960s and early 1970s, when the conflict in Vietnam gave expression to overt Cold War hostilities, American manufacturers of combat planes, engines, and associated systems received a continuing wave of orders. The war's end brought yet another set of changes to combat designs, and the military sector of the aerospace industry expanded its European market through co-production of the General Dynamics F-16 fighter plane. The principal competitors in the market for airliners, Douglas and Boeing, experienced severe financial troubles in the process of developing and selling big new jets. Douglas persevered through a merger with McDonnell; Boeing's foreign sales kept it solvent, and American air transports continued to dominate the global market, although Europe's resurgent aerospace industry continued to grow in strength. As for general aviation, manufacturers enjoyed a booming market through the 1970s until business conditions brought abrupt changes, and a series of mergers and other factors thoroughly restructured the light plane industry.

Missiles and Space Flight

For many years after World War II, the proliferation of nuclear missiles represented a chilling—and hugely expensive—element of the Cold War. At the same time, in many ways such missiles furthered the commercial evolution of microcircuitry and the impact of such technology in modern life. Compact radios, television, and VCRs all benefited from missile research, as did the miniaturized control systems that made autos run more effectively. The Minuteman missile, as one of the first of the nuclear generation, was particularly instrumental in aiding technology transfer to the civil sector.

For the Minuteman, the Autonetics division of North American developed the guidance system, using discrete circuits and transistors. In the early 1960s, the division decided on integrated circuits for the computer installed in Minuteman II, a choice dictated by the far more fluid and variable conditions demanded by a new doctrine of flexible response. From Dallas, Texas Instruments (TI) supplied some 60 percent of the circuits, while Westinghouse delivered the rest. Produced as chips, the first examples sold for $50 to $65 apiece, replacing about one dozen discrete components having equal total value. Pushed by Autonetics, TI eventually improved its production efficiency, reducing the cost to $12 per circuit in 1965. Each Minuteman II computer used about two thousand integrated circuits, and the Air Force eventually deployed several hundred missiles. Volume production meant lower unit prices, opening up new markets in the civil sector. During the late 1960s, the Minuteman effort represented 20 percent of monetary sales for the integrated circuit industry and the trade press remarked that it remained the "largest single consumer of semiconductor microcircuits." The use of these chips in contemporary civilization sprang from the Minuteman undertaking.[1]

At the same time, miniaturized integrated circuitry and related electronic systems continued to evolve with new generations of powerful missiles having awesome accuracy. One such missile was the LGM-118A Peacekeeper, popularly known as the MX missile, originally designed to frustrate the increased accuracy of Soviet ICBMs. Because fixed sites might be totally saturated by enemy missiles, American planners wanted the MX to be shuttled night and day on a rail system to be built under a continuous earthen tunnel. The projected costs reached $30 billion and escalated. The proposals called for various rail-tunnel systems, each running dozens of miles to be built in the West, where (it was argued) scant population offered thousands of empty acres for construction. Governors of the states protested, contending that building such a system would overwhelm existing demands on water supply, unbalance the delicate water tables in the region, and devastate the overall environment.

The Peacekeeper effort evolved as a complicated program, with Boeing responsible for basing the missiles while Martin Marietta was in charge of assembly and testing. Originally, at least 114 Peacekeepers were planned, but the Air Force finally ended up with 50 missiles deployed around Warren AFB in Wyoming. When environmental groups lobbyed heavily against the proposed "rail-garrison mode," their efforts symbolized a new public willingness to resist ready acceptance of national security projects even if they implied some economic benefits through local employment. What appeared to be exorbitant expenses in such systems also ran afoul of increasingly budget-minded Congresses. Consequently, costs were a factor in placing the Peacekeeper in permanent silos, halting secondary plans to randomly transport them among different sites elsewhere in the United States.[2]

Nonetheless, the budget for nuclear missiles remained substantial throughout the 1970s and 1980s, funding U.S. Army Pershing missiles and ground-launched cruise missiles (GLCMs), ordered by the U.S. Air Force. Both were deployed in Europe during the mid-1980s, touching off several rounds of impassioned protests by peace groups overseas. Additional budget costs accompanied the deployment of even more lethal missile subs.

In the late 1980s, the U.S. Navy equipped missile submarines with Trident II weapons, the latest in a series of underwater-launched missiles that began with Polaris in the late 1950s and progressed through the Poseidon and Trident I series. Each subsequent system outclassed its predecessor in speed, range, and destructive power. The Trident I and Trident II both used inertial guidance systems that could make star shots of celestial reference points to stay on course. Trident II also used an enhanced Navstar Global Positioning System, fine-tuning its true position in reference to a grid of satellites overhead for a precision impact on target. All of these weapons, ICBMs, bombers, and a variety of short-range weapons (giving defenders less response time) presented enemy forces with a maddening array of nuclear threats. Defense systems and similar attack weaponry like these were not only hugely expensive for American taxpayers; they became an onerous economic burden for the communist bloc countries and the Soviet regime in particular.

In 1983, President Reagan compounded Soviet problems by committing the United States to a program called the Strategic Defense

Initiative (SDI), popularly known as "Star Wars," after the successful motion picture series. From its inception, the idea generated sharp skepticism from many experts. Basically, SDI represented a response to the Soviet practice of deploying many more long-range nuclear missiles than the United States produced. Rather than increase the number of warheads around the globe, the SDI system proposed to intercept and destroy Soviet missiles as they followed trajectories through space. The "Star Wars" approach reflected the sort of high technology in which the United States excelled and presumed a level of R&D funding that the Soviet regime simply could not match. Although SDI budgets in America totaled $32 billion by 1992 and no successful system ever evolved, the Star Wars effort appears to have been one of several factors in bringing the Soviet leadership to the bargaining table for serious negotiations to control the proliferation of nuclear weaponry.

After two decades of arms limitations talks, a series of crucial agreements between the United States and the Soviet Union coalesced in the late 1980s and early 1990s. During low-keyed but dramatic visits in the 1980s, teams of American and Soviet observers toured high-security military sites in each others' countries to verify the dismantling of short- and medium-range missiles. Further discussions led to a pair of treaties emanating from Strategic Arms Reduction Talks (START I and START II), signed by Russian premier Mikhail Gorbachev and by U.S. Presidents George Bush and Bill Clinton. These historic agreements slashed the nuclear warheads of the two countries by some 60 percent and led to the stand down of planes and subs on permanent alert status. Although the Clinton administration kept some SDI research projects alive, the program received major cuts and lost its priority status in defense planning.[3] This dramatic transition of Cold War attitudes eventually had repercussions for not only aeronautical programs but for American space exploration plans as well.

In terms of manned missions, NASA's dramatic series of Apollo launches to the moon ended with Apollo 17 in 1972. No funding appeared for follow-on missions and no hardware remained in production. Concerned about maintaining the professional edge of launch crews and mission personnel until shuttle operations began, NASA turned to interim ventures. The quartet of Skylab missions during 1973 and 1974 utilized a leftover Saturn S-IVB upper stage, refitted to support three astronauts for limited stays while they performed scientific experiments in earth orbit. The Apollo-Soyuz Test Project (ASTP) operation, which linked Soviet and American manned spacecraft in earth orbit, successfully occurred in 1975. Considering the implicit Cold War rivalries of the two powers in the past, the linkage in space seemed a political paradox, while helping broaden the base for improved relations in the future—not only for space cooperation, but political cooperation as well. None of these undertakings implied further production, although they engaged designers and engineers in the kind of projects NASA hoped to mount in the future: repeated launch and docking of space shuttles with a permanent space station in earth orbit.

The winged shuttle, optimized to launch into space, perform earth-orbital missions, and return for a conventional landing like an aircraft, commanded the lion's share of NASA funding through the 1970s. The shuttle budget came to about $5.2 billion for research, development, and deliveries by the early 1980s. Rockwell became the prime contractor; its Rocketdyne division supplied the craft's main engines; external tanks to supply propellants at launch came from Martin Marietta (produced at NASA's Michoud facility near New Orleans); and Thiokol fabricated the solid-booster rocket motors used during liftoff. NASA's shuttle "fleet" came to only four active-duty vehicles plus one unpowered vehicle for flight tests. With only four operational spacecraft, the comparatively high cost of the shuttle program, like the Apollo-Saturn program, came from exhaustive research and development to make sure that the system would work with a minimum of flight trials. A crucial aspect of international collaboration involved Canada's investment in research, development, and fabrication of remote manipulator arms for the shuttle fleet. The remotely controlled arm, essential to the shuttle's space operations, enabled astronauts to deploy and retrieve bulky hardware and payloads during orbital operations.

NASA planned for the shuttle to support a catalog of earth-orbital missions, using its cargo bay to carry a variety of scientific equipment, including a large, self-contained research module—designed and built by European industry—called Spacelab. Eventually, NASA's plans depended on the shuttle to support the construction of a permanent space station and to provide transportation between it and the earth. After five years of missions, including successful flights with the Spacelab, the loss of *Challenger* and its crew of seven in 1986 marked a tragic reversal of fortune. An investigative board recommended a major reorganization of NASA's own administrative framework as well as its management of contractors.

The shuttle disaster dulled NASA's image as a premier manager of high-technology adventures. Bureaucratically, the agency never again enjoyed quite the level of public support and congressional largesse of earlier years.[4] Nor was this the only functional change in NASA's operations—or, for that matter, activities of the American aerospace industry. Budgetary constraints in the American aerospace community and the maturing of European aerospace capabilities inexorably led to striking international cooperation. This was a trend that had been slowly taking shape through the decades after the end of World War II, gaining much higher visibility during the 1970s and 1980s.

European Aerospace and the NASA Connection

In 1945, with the exception of Great Britain, the European aviation industries of Germany, Italy, and France had been destroyed by the war. As the postwar recovery gained momentum, the British and French aircraft industries both rebounded, eventually followed by those of West Germany and Italy. These four countries, as major scientific and technological centers in Europe, also began developing space technology as well. By the 1970s, organization of the European Space Agency (ESA), plus several successful multinational aircraft programs (such as the aggressive aviation consortium Airbus Industrie), symbolized the robust capability of Europe's space community.

None of this happened in a vacuum. One of the most significant elements in the emergence of European aerospace technology had to do with the philosophical and pragmatic factors in collective European recovery schemes after the war. Against this background, four trends stood out. First was the industrial and scientific cooperation leading to entities such as the Common Market and the European Center for Research in Nuclear Physics. Second was the manufacturing and managerial experience of airframe and engine consortia that had emerged by the 1960s. Third was the evolution of the North Atlantic Treaty Organization (NATO) and the push to develop and deploy high-performance military aircraft of American design to deal with potential Cold War hostilities. Fourth was the series of cautious steps leading to the creation of ESA and an integrated program of space exploration. The development of the unmanned HELIOS project and the manned Spacelab project (as a shuttle payload) afford additional insights into the sources of European aerospace skills and the evolution of significant cooperation with the United States.

In many ways, the most impressive European aerospace program was the Concorde, the result of an Anglo-French collaboration. During the 1950s, when Britain was considering a stable of supersonic bombers, the idea of a supersonic transport gained strength. But the costs were seen to be very high; not only past Britain's means, but more than any single European country could manage. In 1962, after many months of negotiations, the French and British agreed to joint development of a supersonic transport, the Concorde. Following test flights in 1969, an exhaustive test program delayed scheduled service of the Concorde until

1976. Nonetheless, the Concorde experience was highly useful for subsequent joint aircraft programs and for European space programs to come.

During the early 1960s, German production of high-performance fighters like the F-104, followed later in the decade by Italian experience in DC-9 assemblies for Douglas and specialized components for jumbo planes like the DC-10, generated additional areas of expertise. By the time that nascent European space programs were getting under way, Europe not only had the experience of these and the Concorde but also a series of advanced, multinational strike/fighter aircraft like the Jaguar (France, Britain), and the Panavia Tornado (Britain, Italy, West Germany). Moreover, a variety of cooperative experiences in multinational trainers as well as assorted NATO weapons systems added to this useful legacy.[5]

After Sputnik shocked the United States into a heavy investment in space research, the Russian-American space race played out in the spotlight of international attention. Some leaders in Western Europe began to worry about the decline in both prestige and technical abilities unless they developed their own space program. In this atmosphere, European interest in a cooperative development of space technology began to grow. During the years 1962 through 1973, a total of 31 European-built satellites were developed and placed in service by European consortia. Although many were boosted by American launchers, one-third were carried aloft by French or British rockets, including one by the Russian Vostok booster (a French technology satellite, SRET-1, launched April 4, 1972).[6]

A good example of European space progress can be seen in the HELIOS project, using an American Titan/Centaur launch vehicle to carry a German satellite. This was a very complex effort, sending two HELIOS satellites (in 1974 and again in 1976) into solar orbits, closer than had ever been attempted. Although the United States and Germany played the major launch roles, HELIOS represented an industrial effort that drew from subcontractors in Germany, France, Belgium, and Great Britain. The program not only promoted European expertise in spacecraft but also represented an unquestioned achievement in multinational management skills.[7]

Several events led to the Memorandum of Understanding of 1973, which united NASA and Europe in terms of the Spacelab payload for the U.S. shuttle. Contributing factors included NASA budget constraints and American commitment to international space ventures, as well as a European desire to develop the advanced industrial technologies that were inherent in space programs but still outside their immediate reach. By 1975 a mature, unified space community in Western Europe reorganized itself as the European Space Agency (ESA) in Paris. The realignment of programs among ESA's 11 member states also formalized several

R&D establishments, the largest being the European Space Research and Technology Center (ESTEC) in the Netherlands. At about the same time, the German aerospace consortium known as ERNO, in Bremen, was named as the principal industrial firm in the production of Spacelab.[8]

Realizing their lack of experience in manned missions, ESA and ESTEC hired a number of American aerospace technicians to fill specific gaps in Europe's aerospace capability. At ESTEC, the Americans generally served as the immediate deputy to a European in charge of a division or office, imparting knowledge to the European in charge as well as to Europeans who reported to the American deputy. Thus, European executives acquired new depth in aerospace management for manned missions. Experience was sometimes painfully acquired. Changing design requirements of the Orbiter made it the "driver," causing many changes in the concurrent design and development of Spacelab.

These difficulties might have created serious repercussions, except for the determination by all concerned to make the program successful. Considerable credit should go to the Europeans. At both ESTEC and ERNO, managers decided to use English in all correspondence as well as in all staff meetings and minutes. At ERNO, management organized dozens of English-language courses and imported Berlitz experts to make sure everyone—including security officers at the reception desk—became reasonably fluent. A variety of coordinating committees (U.S. and European) also facilitated progress. The Joint Spacelab Working Group, organized at the headquarters level of ESA and NASA, seems to have been particularly effective. Looking back on the experience, American engineers expressed admiration for the high quality and workmanship of European hardware. There was also a strong respect for the Europeans' ability to incorporate so many different national companies, with cultural and linguistic differences, into a workable multinational venture.[9]

Although space shuttle development and operations seemed to garner the most news headlines during the 1970s and 1980s, NASA also launched a long series of scientific payloads into earth orbit, as well as a number of impressive planetary and deep space probes. While the Department of Defense remained the big spender in earth-related payloads (reconnaissance and other satellites), NASA played a significant part, along with the Department of Commerce and the Department of Energy. As for manned missions, the space shuttle and the Spacelab promoted stimulating research by humans in space, especially in terms of studies to realize commercial success from space-based procedures to propagate near-perfect crystals for electronics and to produce other products such as pharmaceuticals. By the close of the 1980s, NASA looked forward to dramatic results from the launch of the unmanned Hubble Space Telescope and deployment of the planned space station *Freedom*, with both programs involving significant international collaboration.

NASA changed from a research organization to an operational agency during the Apollo era, and conseqently it and other federal agencies became major aerospace customers, joining the military services, airlines, and the general aviation community as consumers of aerospace products. While NASA's budget for aerospace hardware did not reach military proportions, its spending remained substantial through the 1970s and 1980s. By 1990, industry sales of space systems amounted to $28.9 billion. Not surprisingly, the Department of Defense outlays claimed the largest portion, at $15.4 billion, but civil orders came to a very respectable share, with NASA at $12.3 billion, the Department of Commerce at $232 million, and the Department of Energy at $79 million. Excluding engines and propulsion units, annual nonmilitary sales represented roughly one-half of the military total since 1975, coming to about $3.4 billion and $6.3 billion, respectively, in 1990. The budget requirements for engines and propulsion units remained essentially equal through the same period, with $1.8 billion in military and $1.3 billion in civil orders for 1990.[10]

Military Aviation

From 1961, the United States gave overt support to the South Vietnamese government, including air operations against the government of North Vietnam. Following the Gulf of Tonkin resolution by Congress in the summer of 1964, U.S. aerial operations escalated to massive proportions, including sorties against targets in North Vietnam, off limits up until that time. The North Vietnamese Air Force, estimated to number about 200 fighters in 1972, restricted its operations to North Vietnamese airspace. USAF strategists gave top priority to the ground war in South Vietnam, Cambodia, and Laos, so that 75 percent of sorties from American air bases in the region were keyed to the support of ground troops in that theater. The principal targets of airpower included areas where enemy troop movements were reported but concentrated on interdiction of lines of supply and communications, armed reconnaissance, and electronic surveillance. In terms of technological advances, the Vietnamese experience had far more to do with electronics and avionics than with the kind of propulsion and aerodynamic progress that marked World War II and Korea. Despite the protracted nature of the war in Vietnam, the impact on the aerospace industry seems to have been more qualitative (electronic advances) than quantitative (mass production of planes). Industry was able to meet operational demands without the radical expansion of plant capacity required in World War II, replacing aircraft in Vietnam that—for the most part—were already in production before the massive American buildup of the mid-1960s.

Helicopters were an exception. Used primarily in Korea for medical evacuation and light transport, the jungle terrain of Vietnam and its characteristics of unconventional warfare rapidly made their advantages—and importance—clear. The U.S. Army relied on helicopters to replace trucks, jeeps, and weapons carriers for fast retaliation after random and geographically diverse enemy attacks, and to make similar attacks against the enemy. American forces deployed 1,000 helicopters into the war zone in 1966, and production drove the numbers up. The Bell AH-1 Huey Cobra, a helicopter gunship, resulted from a crash program to develop a speedy, armored, and well-armed helicopter for escort duties and fire support in Vietnam. Bell Helicopter received well over 1,000 orders for the Huey Cobra alone, which equipped U.S. Marine as well as U.S. Army units. During the course of the war, several thousand helicopters—primarily from Bell, Boeing, and Sikorsky production lines—provided invaluable service in delivering troops and supplies, evacuation, reconnaissance, and other tasks. Due to the high mechanical stress on rotors and rotor machinery, continuous operations took a high toll on helicopters. Moreover, their low-level flights in combat made them unusually vulnerable to enemy ground fire.[11]

Among the many jet fighters deployed in Vietnam, the McDonnell F-4 Phantom II ranked as one of the most versatile and successful. With the F-4 Phantom II, U.S. air forces began operating at twice the speed of sound. The design began in the early 1950s as a successor to McDonnell's FH-1 Phantom but soon changed beyond all recognition. Originally intended as a single-seat fighter, McDonnell's design dramatically altered after the Navy announced requirements for a long-range attack fighter. The new plane, known as the Phantom II, became a two-seater with a large airborne radar in the nose, monitored by the second aviator as weapons operator. A variety of missiles controlled by elaborate electronic systems became its sole air combat protection. With provisions to carry additional bombs and other electronically guided air-to-ground weapons, the plane's size, weight, and complexity mounted. While all of this added up to an expensive mosaic of management systems for engineering and production, it also produced one of the most lethal and legendary combat planes of the postwar era.

By 1961, about the time that some 45 planes had been built and delivered to Navy squadrons for stringent test and training development, the Department of Defense inaugurated a new, unified aircraft designation

The McDonnell Douglas F-4 Phantom was manufactured in large numbers for the U.S. Air Force as well as for the U.S. Navy. Considerable numbers were also sold to foreign air arms; the planes in this production line were built for the Royal Air Force. *Courtesy of the McDonnell Douglas Corporation.*

system for all the military services. Production versions of the Phantom II officially entered service as the F-4B. The Air Force, impressed by the aircraft's capabilities, made the unusual decision to acquire a combat plane from its rival service, and ordered its own version as the F-4C, which began squadron operations during 1963. During the Vietnam War, McDonnell's production lines worked at full tilt, delivering the Phantom II to Air Force as well as Navy and Marine units. Additional models went to air forces elsewhere in the world: Britain's RAF in the mid-1960s; Iran, Israel, and

Korea later in the decade; Australia, West Germany, Japan, Greece, and Turkey in the early 1970s. As the USAF divested itself of some earlier models, Spain also operated the Phantom II, and Egypt, having realigned itself closer to Western Europe and America, received several dozen in 1979. When production finally ended in 1979, more than 5,000 models of the F-4 had been delivered, with slightly more than one-half going to the U.S. Air Force. Because so many remained in foreign service as well as U.S. squadrons through the 1990s, McDonnell and subcontractors also reaped a considerable "after-market" benefit through modification work and upgrading of various systems. Large numbers operated as USAF reconnaissance aircraft and others carried special systems for electronic countermeasures to detect and confuse enemy radar and missile guidance programs. Many F-4 Phantoms were scheduled to remain on the roster of foreign services into the twenty-first century.

Combat experience in Vietnam resulted in changes to the plane's armament. The F-4 Phantom carried Sidewinder as well as radar-guided Sparrow missiles and shot down several dozen enemy aircraft with these weapons. American pilots wanted a gun suitable for close-range work and violent maneuvering, and a 20-mm rapid-fire cannon built by a General Electric subsidiary was incorporated. Production of missiles, cannon, radar, and ancillary equipment all contributed to brisk sales of military aerospace products. Some 950 models of the F-4 were sold to foreign air arms, with an additional 138 built under license in Japan. A rough comparison of prices for fighter aircraft over several decades reflects the value of McDonnell's export market. A World War I fighter cost about $5,000. Twenty to twenty-five years later a World War II combat plane cost approximately $50,000. Fighters of the Korean era climbed to $500,000 apiece, and by the mid-1970s, the price of a fighter plane had escalated to around $5 million. McDonnell's total sales of F-4 Phantom II abroad approximated $4 billion to $5 billion.[12]

Based on the Vietnam experience, the American military forces inaugurated development of postwar combat planes having both missiles and guns for armament; enhanced avionic systems to control the pilot's workload while providing more effective combat situation data; a relatively large wing area; and immense power from the plane's turbofan engines, especially given the rising weight of the airframe and associated systems. In fact, designers laid out the new planes and planned their combat maneuvers to take advantage of their size and extremely high thrust based on a new theory of "energy maneuverability." One of the earliest Air Force planes to emerge from this design process was the McDonnell Douglas F-15A Eagle, making its first flight in 1972. The Navy's new design emerged from Grumman as the F-14 Tomcat. Like the

F-15, the Navy's aircraft was built around two powerful engines but featured variable-sweep wings to enhance takeoff and landing characteristics from carrier decks. A two-man crew was surrounded by a multitude of avionics systems to accommodate the Tomcat's necessary versatility; it functioned in fleet defense and ground attack and as a fighter escort. The F-14 prototype made its initial flight in 1970 and entered squadron service aboard carriers in 1974.

Other specialized types of planes went into production for the services, such as the McDonnell Douglas AV-8A Harrier for the U.S. Marines. Based on the Harriers originally developed in Britain for the RAF, the plane entered American service in 1971. A subsonic aircraft, the Harrier's jet power plant included pairs of exhaust nozzles on either side of the fuselage; swiveled downward, they allowed the plane to make a short takeoff with combat loads and also to make a vertical takeoff and landing—a definite advantage in forward combat areas.[13] The Harrier, in fact, represented a notable international agreement in which British and American industry undertook joint production utilizing certain shared components. However, the number of planes and monetary value of the Harrier program paled in significance with the American aerospace industry's gambit to win the "arms deal of the century" for a new European fighter.

By the early 1970s, the NATO air forces in Europe recognized the need for a new fighter aircraft, especially as a replacement for the Lockheed F-104. Norway, Denmark, Belgium, and Holland played a key role in deciding who would be awarded the contract, based on immediacy of need and numbers of planes likely to be ordered. Four manufacturers emerged as the front runners: Saab of Sweden, Dassault of France, and two American contenders, Northrop and General Dynamics. The Swedish and French candidates seemed problematic, largely because neither nation belonged to NATO. Northrop's proposal, a twin-engine fighter called the F-17, looked promising, although its twin engines implied added costs and maintenance expenses that held little appeal for the smaller, budget-minded forces of Norway, Denmark, and Belgium, in particular. A variety of political and economic factors eventually favored the General Dynamics candidate, the F-16.

The four leading customer countries announced in 1974 that they all agreed to select one plane, representing a guaranteed sale of considerable value. Although the various manufacturers submitted revised proposals, with highly attractive co-production clauses, the General Dynamics offer held the greatest appeal. Indeed, haunted by the controversies over its F-111, bereft of follow-on orders for the swing-wing plane, and confronting its cavernous—and empty—mile-long assembly building in Fort Worth, General Dynamics faced a do-or-die situation over its

F-16. Having won a U.S. Air Force contract, a favorable European decision promised expanded production, lower costs, and higher profits. With greater confidence, the General Dynamics sales team offered a truly appealing package to the Europeans, assisted by an advisory coterie of senior officers from the U.S. Air Force itself.

For one thing, the Americans offered a very favorable co-production package. From the initial production run of around 1,000 aircraft, 650 would go to the U.S. Air Force and 348 to the European consortium; the Europeans would build 10 percent of the American order and no less than 40 percent of the European order. The deal included an attractive lure—15 percent of the work on additional planes ordered by other countries, with deliveries projected to more than 4,000 planes by the 1990s. These figures represented a 58 percent industrial offset, estimated to generate about 25,000 jobs in Europe. All this meant continuing jobs and economic activity along with the transfer of high technology, especially electronics, to the consortium's members. Finally, General Dynamics underbid its competitors with a "not to exceed" price of $6.09 million per plane. Only the Belgian government seemed to hesitate. Just before the Paris Air Show of 1975, the American government arranged a $30 million contract for machine guns from a Belgian firm, and General Dynamics soon announced its success as winner of the sale of the century, estimated at $2.5 billion and bound to increase.

The "not to exceed" price soon became reinterpreted in a blizzard of pettifoggery that included the General Accounting Office. The first F-16 rolled out of the Fort Worth plant in 1976, and the GAO admitted the following year that the additional cost of $4 million for a single F-16 (bringing the per plane cost to $10.8 million in America and $11 million in Europe) had occurred largely due to differing international labor traditions. Denmark had to spend 90 percent of its 1978 military budget on the F-16. Although Belgium received a 70 percent offset, the average offset for the four nations came to about 52 percent. On the whole, the promise of technology transfer seemed unevenly fulfilled; some new companies were created, but little new, innovative technology was introduced to established companies like Fokker. Export sales languished. By the end of the 1970s, European opinion on technology transfer from the F-16 remained very mixed, although exports later multiplied.[14]

While collaboration between U.S. and European manufacturers had been fairly common, collaboration among the different U.S. manufacturers had not. Rising costs and other factors eventually led to the increasing phenomenon of "teaming," which involved cooperative ventures between ostensible U.S. competitors. When the U.S. Navy launched its own low-cost, lightweight fighter program, for instance, Northrop Corporation, a traditional Air Force supplier, realized that it lacked the background to succeed in shipboard operations, and needed additional funding to boot. Northrop joined with McDonnell Douglas to offer a navalized version of one of its own designs, which eventually won the contract under the F-18 designation. The agreement between the two manufacturers stipulated that McDonnell Douglas would develop the F-18 for aircraft carrier operations, with Northrop as prime subcontractor. Northrop was free to develop a land-based version for export. The Navy model became the McDonnell F/A-18, and the veteran Navy supplier took advantage of the plane's twin engines to expand its weapons payload and add high-tech avionics, allowing the plane to sortie as a combined fighter and attack plane. McDonnell also offered this versatile design to land-based air arms, competing with Northrop and touching off a series of heated legal engagements when the latter's version lost out. Although this particular teaming experience left a bitter residue, collaborative ventures became the norm, for they seemed the only way to cope with soaring development costs. The F-18 eventually won several international contracts in addition to service with U.S. Navy and Marine squadrons.[15]

Even though the total production figures for U.S. military aircraft declined during the post-Vietnam era, the costs of modern aircraft, avionics, and engines pushed budgets into the tens of billions. During the 1970s, the total budget for military engines alone was expected to top $10 billion, a market that pitted Pratt & Whitney versus General Electric in what many observers called the "Great Engine War." At the same time, the Department of Defense initiated a new set of procurement procedures that restored an environment of healthy competition for these huge contracts, reversing a tradition of awarding an initial research and development grant to one manufacturer, which then could expect an essentially guaranteed production contract. Over the years, Pratt & Whitney had emerged as the basic supplier for the Air Force and became the exclusive supplier for engines that powered the F-15 and F-16 fighter fleets. Additionally, the company supplied engines for the Navy's F-14. When the Air Force finally changed their procurement practices in 1984, Pratt & Whitney had already shipped over 3,000 engines worth $10 billion.

In the early 1970s, when Pratt & Whitney had first wrapped up engine contracts for the F-14, F-15, and F-16, General Electric won the contract to supply engines for the Rockwell B-1 bomber but ran into a stone wall when the Carter administration canceled the bomber, leaving GE with a promising engine having no clear market. GE nevertheless continued to develop the engine, known as the F101. Later in the decade the company received Air Force and congressional support during efforts to revive competition in fighter plane engines. Around this time, the Pratt & Whitney F100 was experiencing delivery problems resulting from strikes by two important vendors, reinforcing Air Force desire to have an alternative, competitive engine manufacturer. When President Reagan revived the B-1 bomber in 1981, prospects for the GE fighter plane engine unit, now called the F101 DFE, came to life with a vengeance. Pratt & Whitney fought its newly reinvigorated rival with a passion. A flurry of congressional hearings and debates ensued during the spring of 1983, with aggressive congressional debate and corporate offensives from Connecticut-based Pratt & Whitney, but the Air Force and its allies in Congress persevered in support of competitive engine procurement. Early in 1984, a new Department of Defense contract divided the production awards for funding year 1985, with 25 percent (40 engines) given to Pratt & Whitney and 75 percent (120 engines) to GE, a decision ratified that spring by the General Accounting Office after a careful inspection of the contract awards. Consequently, GE not only equipped B-1 squadrons but also won contracts from the Navy to supply engines for the Grumman F-14, as well as from General Dynamics to supply engines for the F-16 orders from Israel and Turkey.

Pratt & Whitney gained little from the great engine war. On the other hand, after hanging on grimly through a decade of hostile lobbying, GE's engine division received a new lease on life. Success also contributed to the learning curve and production economics for a next-generation commercial air transport engine from GE, the CFM-56. Finally, the Air Force and the Department of Defense won praise for their commitment to a more competitive environment, a stance that received crucial support from sympathetic leaders in Congress and set the stage for more competition in the future.[16]

The Airliners

The aeronautics industry's personalities and corporate structures continued to change. Within a decade or so after 1945, most of the prewar industry pioneers had been supplanted by others. William Boeing had already relinquished command to others in the 1930s and died in 1956; a number of his successors had either died or retired. Similar changes occurred at the Douglas Company; at age 65, Donald Douglas took the position of chairman of the board in 1957. Donald W. Douglas, Jr., became president and soon installed his own corporate team. To some observers, Douglas Aircraft would never be the same in spirit, nor as successful as a company.[17] For several years, annual reports noted impressive revenues, although a variety of problems began to assume dangerous proportions.

By all appearances in the mid-1960s, Douglas seemed to be in an enviable position. The DC-8 and DC-9 airliners enjoyed strong sales, production of Navy combat planes continued, and the company held major contracts for missiles and space hardware. But Douglas began to run short of money to keep production lines operating. With a loss of $16 million by the end of the third quarter of 1966 and rumors of a deficit as high as four times that amount by year's end, banks and creditors declined to carry the company's indebtedness. Reviewing the disaster, analysts realized that Douglas had become a victim of its own success. The company's market offensive to sell airliners succeeded so well that the flood of orders led to hasty efforts to boost production; costs escalated. On top of this, workers put in massive chunks of expensive overtime, and procedures to train new production workers meant added costs. Delays in deliveries caused airline customers to hold back payments, and heavy orders resulting from the Vietnam War caused a shortage of aerospace components throughout the industry, triggering further delivery delays. Finally, the company's accounting procedures were judged to be antiquated and had failed to anticipate the cash flow problems or deal with them once they showed up. Douglas called in a team of financiers for advice; the recommendation was to find a merger partner with deep pockets.

For all its problems, Douglas was still a major prize. A half-dozen corporations made serious bids before year's end, including General Dynamics, North American, Fairchild-Hiller, Martin Marietta, McDonnell, and Signal Oil and Gas. In the end, McDonnell's offer succeeded. James McDonnell and Donald Douglas, Sr., the respective patriarchs, shared a Scottish ancestry. More importantly, the two had conducted talks about partnership since the 1950s, when McDonnell began serious thinking about commercial postwar ventures to balance its own successful military production contracts. McDonnell himself recalled discussions between the two men about historic merger patterns in the auto industry that would soon be repeated in the aviation business. Although nothing came of their talks at the time, McDonnell assigned one of his vice presidents, Robert Hage, to keep an eye on commercial products and especially Douglas Aircraft. An ex–Air Force major, Hage was first an executive at Boeing and then joined McDonnell in 1958; his observer role intensified in 1963. "I remember about every three months [James] McDonnell would call me in and want to know just where Douglas Aircraft Company stood," he recalled. "And in the fall of 1966, when the Douglas Company began to experience financial difficulties, we were right there."[18]

McDonnell kept outbidding his rivals, eventually buying 1.5 million Douglas shares for nearly $69 million. From its headquarters in St.

Louis, the McDonnell Company employed 45,000 people, far fewer than the Douglas payroll of 80,000, but the St. Louis manufacturer had substantial profits as well as a rock-solid credit rating from its work for the National Aeronautics and Space Administration and from its phenomenally successful line of Phantom F-4 fighters. McDonnell's years of studying Douglas Aircraft contributed to polished legal briefs for the Justice Department before the merger was officially sanctioned in 1967, and 10 banks supplied a $300 million revolving credit fund. In 1968, McDonnell Douglas Corporation reported earnings of $94 million on sales of $3.6 billion. James McDonnell had long desired a better balance between government and commercial business; the 1968 corporate report indicated success, with roughly 47 percent of income coming from commercial aircraft, 35 percent from military sales, 15 percent from spacecraft and missiles, and 3 percent from electronics, automation, and other services. With confidence, the merged company decided to go ahead with a wide-body airliner, the DC-10. Lockheed and Boeing also had wide-body designs in the works, and a European consortium, Airbus Industrie, announced its own wide-body proposal at about the same time. Obviously, keen competition was in store.[19]

The first step toward bigger jets came when Douglas increased dramatically the number of passenger seats in the DC-8. Anxious to recoup its traditional dominance in the airliner manufacturing industry, Douglas set out to surpass upstart Boeing in this field. Attempting to save time and money, Douglas decided to stretch its basic DC-8 into a new "Super Sixty" series. The first of these, the DC-8-61, took to the air in 1965. The plane's fuselage included an added segment to make it nearly 37 feet longer than its predecessor, and the new version seated up to 259 passengers, compared to a maximum of 189 for early DC-8 models. The seating capacity was unprecedented for the time, and the travelers' first glimpse of the single cabin aisle, which seemed to stretch forever, was impressive. In lengthening the DC-8, Douglas realized an unanticipated advantage. The wing sweep and its location on the fuselage permitted the fuselage of the "dash-sixty" series to use the basic DC-8 wing manufacturing jigs and landing gear hardware. On the other hand, when Boeing engineers sat down to consider similar changes to the 707, they discovered that an elongated passenger cabin, given the Boeing's wing sweep and landing gear location, would have the tail scraping the runway each time the plane rotated into the necessary angle for takeoff. The Boeing people realized that modifications to a stretched 707 would require too much redesign and engineering to be cost-effective. The logic for building a new—and much bigger—airplane became inescapable. Boeing decided to enlarge the fuselage as well as stretch it, leading to the generation of "jumbo jets."[20]

Boeing entered its 747 design studies with momentum carried over from the company's proposals for an oversized military cargo plane, the project that went to Lockheed as the C-5A, and the legacy of the military design studies obviously contributed to rapid progress of the new plane's design. Boeing also introduced computerized design analysis on a large scale. Pan Am, impressed by the new plane's specifications, placed the first orders (for 25 aircraft) in April of 1966, just a few months after Boeing started work on the design. At an estimated cost of $15 million to $18 million per plane for deliveries in the late 1960s, a Pan Am executive recalled that the airline was required to sign "a prepayment scheme far more exacting than any previous experience." With 25 planes on order, Pan Am committed to pay Boeing's down payment of $250 million even before the FAA granted certification. Boeing itself had to commit over $2 billion in development and production costs with no guarantee of success. Indeed, it was not clear if either the builder or its customer would survive.[21]

The size of the 747 precluded the use of existing plant space, so Boeing had to invest in new facilities of commensurate dimensions. In January 1966, the company began clearing 780 acres of timbered hills and absolutely undeveloped land, not far from Everett, Washington. The facilities in Everett were more of an assembly complex than a manufacturing plant. The only major components built by Boeing were the plane's huge wings and the 33-foot-long forward fuselage section, which housed the flight deck; as much as 65 percent of the 747 originated with subcontractors. When the program matured, as much as $2.1 billion in subcontract agreements went to large and small firms, with 20,000 different companies involved. The subcontractors operated in all 50 states and spilled over to include several countries in Europe and the Far East. The first ceremonial rollout of the 747 came during the fall of 1968, followed by production deliveries to airlines by the end of 1969. The meaning of the learning curve became clear in the manufacture of the Boeing 747. In 1969, Boeing assigned 25,000 employees to roll out seven of the big airliners each month. In 1980, 11,000 workers accomplished the same job.[22]

Boeing evolved as a true aerospace giant, with about 142,400 employees during 1968, when sales hit a record $3.3 billion. Things were so good, some analysts accused Boeing of resting on its oars. Locals in Seattle, pleased with Boeing's perennially reliable payroll record, comfortably referred to the company as "The Lazy B." The winds of change arrived overnight. NASA had already begun moves to crank down the lunar exploration program, and the national economy had also begun to turn down, cutting demands for airliners. Boeing's own military production lines were occupied with small numbers of specialized aircraft, some modification work, and long rows of silent machinery. Boeing had high

In a cavernous assembly building near Seattle, Washington, Boeing employees integrate all the components to complete a 747 jumbo jet. Between 1970 and 1995, Boeing sold 1,000 examples of the plane, for a total of $148.1 billion, of which $115 billion came from foreign customers. *Courtesy of the Boeing Company.*

hopes for work on the supersonic transport (SST), but Congress halted that project in 1971. During the same year, Boeing's employment sank to 56,300. Near Interstate Highway 5, a principal urban artery, some disenchanted citizen erected a prominent sign, soon featured in news photos across the nation, with a cynical request: "Will the last person leaving Seattle turn out the lights."

Boeing's tribulations in 1972 sparked a determined effort to diversify in order to avoid the seemingly inevitable cyclic traumas of aerospace manufacturing. The company became involved with light rail transportation, hydrofoils, computer services, urban planning, desalination plants, and a roster of other potential profit ventures. Still, the core of the company involved airplanes. Eventually, the economy picked up and Boeing sold more big 747 transports, along with a smaller plane that nearly failed—the 737.

In the meantime, foreign markets—unhampered by the three-man rule—began to improve. In 1981, a presidential panel in the United States, formed to consider two- versus three-person crews in larger versions of the DC-9, endorsed the two-person concept, adding a domestic market option for the Boeing 737 as well. By 1985, the 737 had passed the DC-9 in sales and led the world in commercial orders. By 1990, the 737 passed the 727 in terms of total delivered units, becoming Boeing's new cash cow. Many factors helped account for record sales, but the early decision for a six-abreast passenger arrangement gave a seat-mile cost that many economy-minded operators could hardly resist.[23]

During the 1970s, the aerospace industry progressed through various changes in engineering procedures for airframe and power plant design. The computer technology of the 1960s had evolved considerably, resulting in unusually powerful new tools in the decades that followed. One of the most valuable functions performed by computers involved computational fluid dynamics, in which a supercomputer generated essential data about wing shapes at very high speeds and altitudes, for example, that could not possibly be duplicated in a wind tunnel.[24] For power plants, the powerful new computational resources helped in designing a new family of high-bypass ratio turbofan engines. Capable of generating very high thrust levels of 40,000 to 50,000 lbs they not only powered advanced fighters like the F-14 and F-15 but also appeared on big new wide-body airliners like the Boeing 747, McDonnell Douglas DC-10, and Lockheed Tristar. Offering high-thrust, economic fuel consumption and quieter operation, the high-bypass ratio turbofans became immensely popular with the airlines as well as noise-conscious communities proximate to airports. Their reliability and impressive power also cleared the way for yet another new generation of airliners of considerable size that needed only two such engines to power them.

Early American examples of this type were the Boeing 757 and 767. The medium-range 757 was initially intended as an eventual replacement for the 727, and when design began in 1978, it retained the same fuselage cross section. Boeing wanted to profit from this standard shape, which dated from the 707 and appeared in the 737, but market a plane with modern engines and aerodynamics. The original plan to graft the flight deck from the 727 to the new plane was eventually dropped, so that the 757 and 767 possessed commonality permitting pilots to get a single rating for either plane. The 767 was seen as a much different commercial type, having a wide-body cabin with two aisles. Its size and medium to long range evolved from the need to compete with new European transports under development by Airbus Industrie. Boeing engineers designed the plane for extensive use of advanced materials, from metal alloys to synthetic composites, as well as electronic flight information systems (EFISs) using computerized visual monitors (cathode ray tube, or CRT displays) in place of conventional instrument dials and gauges. The changing EFIS/CRT displays appearing on the instrument panel gave rise to the descriptive phrase "glass cockpit," a feature that had been marketed earlier in planes developed by Airbus. Deliveries of both Boeing airliners went to customers toward the end of 1982; the 767 began scheduled flights in December, followed by the 757 in January 1983.[25]

General Aviation

During the late 1950s, Piper's studies of potential markets pointed toward a low-cost family tourer featuring low maintenance and versatility to serve basic business flying travelers. The company's factory at Lock Haven, Pennsylvania, had used up nearly all of its space for expansion, and repeated flooding of the west branch of the Susquehanna River represented a continuing threat to production lines. Piper decided on a new production site at Vero Beach, Florida, taking advantage of low land cost, available labor, and sunny weather for frequent development flights and student flight training. Piper designed a new, all-metal plane, the Cherokee, and built the Florida plant to accommodate its manufacture as a mass-production aircraft in several different versions, including trainers. The first planes rolled out the door in 1960, and within five years, Piper delivered 6,000 units.

The Vero Beach operation represented a major shift in Piper's philosophy, not only moving away from its traditional, fabric-covered designs but also producing an all-metal plane designed for the low end of the market. The PA-28-140 Cherokee mounted a 140-hp Lycoming engine; the PA-28-150 meant 150 hp, and so on. The latter still sold for under $10,000, the sort of pricing that made the Cherokee series so successful and made the Vero Beach operation seem like the wave of the future. Indeed, the Cherokee set the pattern for nearly all the piston-engine singles that Piper churned out through the 1980s.[26]

Piper's experience seemed to set the tone for the light plane industry during the 1970s and 1980s. Production figures set new records and the industry reported comfortable profits. Japan, France, and Britain entered the market with new designs for advanced corporate turboprops and jets. With few exceptions (such as a fling with pressurized single-engine models and Cessna's low-price corporate twin jet, the Citation), the planes that American manufacturers turned out were essentially built to the same patterns established in the first two postwar decades. This conservative formula succeeded in the face of adversities that pestered the national economy—zooming inflation, a dramatic fuel shortage, and increasing constraints on general aviation flights around major airports. In 1978, the light plane industry sold nearly 17,000 planes.[27]

As a result of these trends and other related factors, the character of general aviation and the light plane industry experienced dramatic change. In the case of Piper, its success made it the target of several corporate takeovers in the merger-and-acquisition craze that swept through the American business community in the 1970s and 1980s. The generally prosperous 1960s had been good for the Piper Aircraft Corporation, generating record sales (up by 20 percent, to $106 million for 1968–69 alone) and a healthy balance sheet. Piper was ripe for a takeover by one of the conglomerates that had become active during the decade. Early in 1969, Chris-Craft Industries tried just that. A well-known manufacturer of recreational boats, Chris-Craft had branched out into synthetics, chemicals, and broadcasting. The Piper family fought back, wanting to maintain managerial autonomy, but eventually realized that they would have to find a merger partner that could outbid Chris-Craft and also promise operational independence. Bangor Punta, a Fortune 500 conglomerate that turned out such diverse products as sailboats, motor homes, law enforcement hardware, environmental control systems, and more, seemed like a good candidate, and late in the summer of 1969, Bangor Punta had indeed beaten out Chris-Craft. Piper Aircraft Corporation continued to produce airplanes but no longer operated as a family affair. W.T. Piper, Sr., died in 1970, and the Piper family eventually vacated their positions as company officers.

In the meantime, Piper's activities had expanded overseas, with arrangements in Latin America for the assembly of aircraft from components shipped from the United States. During 1974–75, Piper entered into a more comprehensive agreement with Brazil for the indigenous production of Piper designs under the direction of a government manufacturer, Embraer. Brazil wanted to develop its own technical, manufacturing, managerial, and marketing capabilities in general aviation, as well as to control imports that drained Brazilian finances. As production of Piper aircraft got underway during 1975–76, the government imposed a 50 percent tax on imported planes, creating a protected market within the country. By 1983, Brazil was producing six single-engine types and two twin-engine types, based on Piper designs but marketed under Embraer designations and carrying distinctive Brazilian names like the "Carioca" model. In addition, Embraer produced an agricultural plane and developed its own line of larger twin-engine designs, including the "Bandeirante," an 18-passenger commuter liner that sold well in the United States and in other foreign markets. A wrenching series of takeovers by several different financial groups left Piper badly battered by the late 1980s. A savior appeared in the person of M. Stuart Millar, a World War II fighter pilot with a postwar record of multimillion-dollar investments and a man who still loved flying and the lure of Piper's historic name. He replaced most of the old management, focused on light piston engine singles and twins (markets abandoned by other major manufacturers), and aggressively countered outstanding liability suits against Piper. The company began a slow march toward recovery.[28]

Cessna had begun overseas operations during the 1960s, establishing fabrication sites in Argentina and France. The French venture eventually assumed greater autonomy, marketing planes in Europe under the designation Reims Aviation, taken from the city of its location. The Reims organization primarily produced standard Cessna airplanes, although some designs were modified for specialized needs in the European and African markets. In 1983, the Reims unit played a role in developing a new twin-engine design for utility and rugged bush-flying operations. For its larger twin-engine planes and jets that sold overseas, Cessna set up a marketing operation based in the United States.

Beech, normally a conservative organization, made some interesting moves to remain a major player in the light plane industry. Realizing the value of having a larger corporate jet to offer companies wanting to upgrade from Beech's own King Air pressurized turboprop, the manufacturer nonetheless paled at the expenses associated with research, development, and tooling. Instead, the company went international in 1970, signing an agreement to import the impressive Hawker Siddeley 125 from Britain, reequipping the plane with American avionics and customized interiors suited to the North American market. The arrangement lasted for five years and ended amicably enough when Beech decided that it needed more authority in building and marketing its own product. During the 1980s, the Japanese aerospace giant, Mitsubishi, began to retreat from the general aviation business, and Beech purchased its business jet design in 1985. Known as the Diamond, the Japanese plane had been a sales disappointment for Mitsubishi. Beech reworked it as the Model 400, updating its cockpit design and other internal systems, revamping its interiors, and turning it into a market success. Although Beech continued as a leader in selling big corporate turboprops like the King Air, the company also realized that its basic design, originating in the 1960s, would eventually need replacement. The company launched a truly radical alternative, called the Starship, planned and developed in collaboration with the cutting edge designer Burt Rutan. Its novel shape and innovative composite construction made international aviation headlines. After extensive flight development in the late 1980s, deliveries began in the 1990s.

Beech's impressive statistical records and balance sheets attracted many merger-minded companies. By the end of the 1970s, Beech recorded sales of $500 million and became increasingly nervous about an unfriendly takeover. The company sought its own partner, eventually coming to an agreement with Raytheon in 1980. Raytheon's principal activity in electronics gave Beech new opportunities in technological capability and product development, while Beech's success in the civil sector of aviation added balance to Raytheon's military aerospace business. Mooney, which had also been subjected to several different owners, finally reached security through an international arrangement with the French company Euralair, an air charter and cargo operator. The Cessna Corporation also hoped for a white knight and in 1985 agreed to an appealing offer from the aerospace giant General Dynamics. The Grumman Gulfstream corporate operation became an independent company as Gulfstream Aerospace, was owned by Chrysler for a period in the 1980s, and was bought out by a new investor group early in the 1990s. Gulfstream closed its turboprop line after developing an even larger, more luxurious corporate jet, the Gulfstream G-III, superseded by the G-IV and G-V. Gulfstream successfully sold its very expensive planes while the overall market for smaller, less costly general aviation planes languished. As one aviation writer commented, "While builders of small aircraft have had to seek new niches and markets in order to survive . . . , others simply have made the big and fast bigger and faster."[29] For Gulfstream, market specialization worked very well.

Like military aviation and space exploration, the light plane industry was not immune to political crosswinds and international realities. Forces long in play brought dramatic changes to general aviation operations and to the light plane industry during the 1970s and 1980s. Since the late 1960s, airspace congestion around major airports had been a major concern of the Federal Aviation Agency. Beginning in the 1970s, new regulations expanded air traffic control and required additional electronic apparatus. While losing some freedom of movement, general aviation activities became safer, avionics manufacturers sold more units—and aircraft prices went up as more avionics equipment appeared in the cockpit. Other changes were brought about by the decision, in 1973, of the Organization of Petroleum Exporting Countries (OPEC) to cut production and hike crude oil prices. A rationing system constrained general aviation, and continued high prices for fuel had a negative effect that persisted into the 1980s. A rash of expensive court actions during the mid-1970s had disastrous consequences for the industry. Legal suits against manufacturers for alleged crash-related deficiencies snowballed and were often settled at the expense of the manufacturers. The industry argued that pilot proficiency was often a factor in accidents and noted that manufacturers had no way of insuring that adequate repair and maintenance procedures took place, especially for older planes that had passed through the hands of several owners. Settlement costs from these liability suits ran into the hundreds of millions and industry executives stated that liability insurance added 20 percent to 30 percent to the cost of each new plane they sold.

Total sales for single-engine piston light planes fell off drastically during the early 1980s. High fuel prices and climbing interest rates were partly to blame, as was the earlier success of the industry itself, when annual production records helped create a considerable supply of now-used aircraft in good condition. All of these factors interacted with the dramatically higher prices of new planes, which manufacturers blamed on the spiraling costs of product liability insurance. In any case, many buyers found it more reasonable to buy from the used plane market, not to buy at all, or perhaps acquire a larger corporate plane and plan to increase its utilization. Consequently, Cessna suspended production of all singles in 1986 with the exception of the Caravan I, a capacious, high-wing turboprop marketed as a light cargo plane and widely used by the small-package delivery operators. Beech maintained an assembly line for slow but steady sales of an updated Bonanza F33, with a conventional tail, phasing out the classic V-tailed Bonanza and two other singles. Piper's deliveries of piston singles essentially halted as a function of its ongoing corporate reorganizations during the late 1980s. Mooney sold a modest number of singles in a market that itself remained modest, and a handful of specialized manufacturers hung on, like the Maule Company in Moultrie, Georgia, which built tough little planes for demanding operations in the Canadian and Alaskan bush. But the eye-catching numbers and crowded production lines of Beech, Cessna, and Piper faded into history.[30]

Budget constraints, a thaw in the Cold War, and international collaboration for many aerospace programs meant shifting priorities and industrial alliances. Although the armed forces continued to deploy new missile systems, highly expensive ventures like the MX system and the SDI program ran into serious opposition from both Congress, which objected to escalating defense costs, and the public. At the same time, Strategic Arms Limitation Talks ushered in a new era in superpower relations. Abroad, the European aerospace industry fielded a variety of consortia to compete with the United States in civil aircraft, military aviation, and space technology. Political and budgetary consideration led to significant cooperation involving NASA and the European Space Agency.

The war in Vietnam resulted in another new generation of fighters and other aircraft, with much higher price tags due to advanced technologies, especially in terms of avionics and engines. For airframe and power plant manufacturers alike, competition for billion-dollar orders intensified; international sales and international production agreements became requisites for profitability. For some manufacturers, like Douglas, merger became a means for survival. Boeing and others found themselves forced to take huge financial gambles in the development of new aircraft. International sales in such instances became crucial and essentially rescued floundering programs such as the 737 airliner. International influences were evident in the operations of general aviation manufacturers as well. The most significant factor for general aviation had to do with its production records. "While not as noteworthy for being a technological hot-bed of activity as the 1940s and 1950s," one analyst concluded, "the 1970s was a decade of immense production (no small feat in itself). . . ."[31]

CHAPTER 8
Global Perspectives 1980s/1990s

THE TECHNOLOGICAL leadership represented by the flying boats and transports of the 1930s, the production feats of World War II, postwar exports of thousands of advanced military planes, and success in the arena of space exploration had for decades given the American aerospace industry a sense of nearly absolute dominance in the global marketplace. For all that, a number of postwar arrangements symbolized a significant shift in the manner of conducting business. Complex overseas production programs like the F-16 (1970s) and collaborative efforts like the Spacelab (1970s–1980s) became examples of a new era of international collaboration. Entering the 1990s, overseas agreements became especially comprehensive in the case of the airline manufacturers, and general aviation had moved overseas for market advantages.

As the trend toward international arrangements gathered momentum, global politics forced another critical adjustment in the nature of the aerospace industry. The collapse of the Soviet Union during 1991–92 and its replacement by the Confederation of Independent States (CIS) had serious ramifications for U.S. aerospace research and production. Many observers, acknowledging a broad spectrum of interrelated factors in the dissolution of the U.S.S.R., felt that the Soviet decline stemmed in large measure from the economic strain of attempting to match NATO's air, ground, and naval capability, especially the levels of sophisticated American aerospace technology. The Soviet collapse was punctuated by the outbreak of the Gulf War, which was accompanied by highly publicized video images of advanced American military hardware in action. Paradoxically, despite the striking success of American aerospace technology, the industry entered an era of dramatic restructuring and downsizing, essentially the result of declining Soviet threat.

Airliners: The Foreign Connection

In its *Newsletter* of July 1988, the Aerospace Industries Association's lead article was headlined: "Internationalization: What Is It? Why Is It? Will It Go Away?" The article noted that the U.S. aerospace industry was going through a "major period of change, adjusting to the new global market environment. . . . " The article went on to summarize the plight of American manufacturers, which were finding that the development costs of a new engine or airliner could exceed the net worth of the company. Given the intensity of competition among American aerospace manufacturers, striking a deal with foreign companies was often the most logical option. Arrangements in the past had included co-production and licensing agreements, but the AIA emphasized that current trends were notably different: "joint ventures and collaborative arrangements in the design, production, and marketing of aerospace products and systems."[1]

The significance of all this lay in the former preeminence of the American aerospace industry. Since 1945, America had been the undisputed world leader in a technology—aerospace—that was considered one of the principal indicators of national and international strength. As America lost its lead in autos and other manufactures to foreign competitors, aerospace remained virtually alone in posting a positive trade balance in the 1970s and 1980s. America especially dominated civil airliner production, outdistancing all foreign competitors in the years after World War II. For example, of all jet airliners delivered between 1958 and 1985, 87 percent came from American manufacturers.[2]

Meanwhile, the pattern of airliner manufacturing as practiced by American firms had begun to change. Until the Second World War, it remained the habit of most manufacturers to produce nearly everything in-house, except for engines, radios, and a few other standardized accessories. By the 1950s, subcontractors produced as much as 40 percent of a typical, propeller-driven airliner, and by the time the Boeing 747 entered service in 1969, subcontractors accounted for as much as 70 percent of the airframe's components. There were a number of factors involved, including complexities of airliner production and the growing size of component parts, which made it convenient to shift work to other vendors. Although most of the subcontractors were American, a growing number of overseas firms won contracts, a trend that began in the 1960s. Boeing was one of the earliest to subcontract overseas for airliner components, beginning with a Japanese firm to supply gear boxes and mechanical assemblies for the Boeing 707 in the early 1960s. When the 727 tri-jet entered service in 1964, it carried these plus honeycomb core assemblies, in addition to galleys and lavatories, all made in Japan, and the pattern rapidly expanded to include foreign manufacturers elsewhere. Between 1978 and 1983, Boeing's airliner production line relied on 170 foreign companies from 13 different countries, which supplied everything from major structures to machine tooling and engineering support.[3]

Japanese acquisition of American manufacturing techniques and management began as early as 1955, when North American granted a contract for licensed manufacture of its F-86F fighter. By the 1980s, companies like Mitsubishi were rapidly expanding their capabilities. Operating under coproduction agreements or a subcontract, the Japanese used exactly the same procedures established by the United States. For the most part, manufacturing plants were equipped with indigenous Japanese machinery, but exact replication of some components required the same tooling as used in the States. For production of the Boeing 767, the Japanese invested in a variety of complex American manufacturing paraphernalia, like large brake presses, section rollers, large hydraulic presses, electrochemical milling machines, electron beam welders, autoclaves, and test/control equipment.

Japanese production of large components for American aircraft became quite diversified. As part of a co-production effort with McDonnell Douglas, Mitsubishi Aircraft built fuselage sections for the Japanese version of the F-15 fighter, and also completed final assembly and flight tests before delivery to Japanese squadrons. For the Japanese-built version of the Lockheed P-3C patrol plane, built by Kawasaki, Mitsubishi supplied forward and aft fuselage sections. Airliner business was similarly diverse, with components shipped across the Pacific to California: tail cones for the Douglas DC-10, inboard flaps for Boeing's 747SP, and fuselage sections for the 767.[4]

Given the extent of subcontracting, and the fact that foreign manufacturers were supplying major portions of the airframe itself, it was a short step from subcontracting to shared-risk production. In this respect, it was Boeing's American rival, Douglas Aircraft, that set the pace. The DC-9 was announced by Douglas in the spring of 1963, following various studies over the preceding four years. It was a gamble; there were no firm orders when Douglas decided to go ahead with the project, and the company was already having financial difficulties in producing the larger, four-engined DC-8. Since the money problem had been a nagging issue, Douglas took a new direction in financing the DC-9. The company persuaded some 20 equipment and component manufacturers to share the costs. In this respect, Douglas pioneered the concept of shared-risk production. Although the principal partners were American, there was a notable exception—de Haviland of Canada. In 1965, the Canadian firm agreed to assume a considerable portion of the DC-9's cost by producing the wing, betting that its investment in tooling, materials, and personnel would eventually pay off through future sales of the airplane. The Canadians—like the American risk partners—had confidence that Douglas's position in the air transport industry would inevitably bring substantial orders. They were right, because Delta soon placed orders and options for 30 aircraft, and other airlines quickly lined up.[5] After merging with McDonnell in 1967, the basic design was renamed the MD-80, and the aircraft continued production in several different versions.

In the meantime, Douglas had expanded its foreign partnership in 1966 through a subcontract with the Italian firm of Aerfer, which became part of the Aeritalia organization in 1969. This ongoing work involved fabrication of DC-9 fuselage panels amounting to 20 percent of the airframe value. Aeritalia's performance not only won praise from the DC-9 management team but also led to even more significant arrangements with the American aerospace industry.

The initial experience of collaboration with foreign contractors worked so well that McDonnell Douglas decided to expand on it when negotiating with suppliers for the DC-10 wide-body tri-jet. When the production decision was made, the company hosted a Supplier Symposium in Long Beach, California, attended by some 700 suppliers and representatives. Eventually, more than 2,200 suppliers won contracts. In effect, the DC-10 "production line" operated all over the United States and four foreign countries. Mitsubishi in Nagoya, Japan, made the entire tail cone assembly; Dowty Rotol in Gloucester, England, supplied the nose landing gear. Additional landing gear components came from Abex Industries of Canada, in Montreal, and wing skin forming was accomplished by a Toronto firm called Metal Improvement Company.

From Aeritalia, in Naples, Italy, McDonnell Douglas received the vertical stabilizer for the DC-10, for which the Italian firm had complete design as well as manufacturing responsibility. Eventually, as a subcontractor to General Dynamics, Aeritalia also produced DC-10 fuselage panels, giving the Italians a not inconsiderable role in the DC-10. Finally, the massive wings were produced at a newly organized McDonnell Douglas subsidiary in Toronto. The risk-sharing arrangement of the DC-9 did not necessarily carry over into all the details of DC-10 production, but the growing roster of foreign suppliers is notable.[6] Shared-risk partnership did not mean that the foreign corporation became an exclusive partner of the American producer. Aeritalia cheerfully produced for Boeing as well as Douglas.

Starting in 1971, Aeritalia began a long-term relationship with Boeing, sending a team to Seattle to collaborate on the design of a short-haul airliner with STOL (short takeoff and landing) capability. This venture had a rocky history, and the Italians had already gone home by the time design studies ended. But the association led to a significant arrangement for production of components for Boeing's new Model 767 wide-body twin-jet. In 1978, Aeritalia took on a 15 percent share of the airframe design and manufacture, along with a similar proportion of the financial risk.

Specifically, the Aeritalia commitment included all moving surfaces of the wing and tail: slats, flaps, ailerons, elevators, and rudder, plus the vertical fin itself and the airliner's nose cap radome. The most significant factor in the agreement was the fact that all these components were to be fabricated from advanced composite materials. The Boeing 767 pioneered extensive use of composites in secondary structures such as moving surfaces, and use of some 3,400 pounds of composites in the airliner saved 1,250 pounds of structural weight. As producer of the lion's share of the composite materials, Aeritalia had to accommodate a large infusion of advanced equipment and techniques. Boeing's strict specifications taught the company new disciplines, said Aeritalia's general manager. Moreover, the experience gave the Italians the sort of capability and expertise to successfully negotiate other new ventures, such as the ATR-42 feeder turboprop, produced with France's Aerospatiale and other European partners.[7]

Collaboration invariably meant adjustments on both sides, allowing for national differences and inevitable snarls. For the Lockheed Company, international collaboration brought the corporation to the brink of disaster. As it unfolded during the 1970s, the saga of the Lockheed L-1011 Tristar airliner and the Rolls-Royce RB.211 turbofan engine was uniquely complex. Lockheed chose the engine because of its low noise, fuel efficiency, and light weight, which were made even more appealing by Rolls-Royce's competitive cost proposals. Because the airliner and the engine were both under development at the same time, the plane and its power plants could be more advantageously tailored to each other. Given the experimental nature of the RB.211 engine, Lockheed and Rolls-Royce were dependent on each other to the extent that the Tristar constituted a shared risk program.

Although the prototype Tristar made its maiden flight in 1970, development troubles with the Rolls-Royce engines nearly torpedoed the entire program. Lockheed already faced huge financial troubles from its C-5 transport contract and the delays involving Rolls-Royce's RB.211 engine threatened the entire company, since impatient airline customers might cancel orders for the L-1011 Tristar and stop payments. Starved for cash, Lockheed would have no way to meet its credit obligations. During 1971, increasingly frantic negotiations between Lockheed and Rolls-Royce finally drew in top government officials of both nations. Prime Minister Heath and President Nixon received daily briefings. The British government finally decided to refinance Rolls-Royce, and the Nixon administration secured a guaranteed loan of $250 million for Lockheed. It was a close call. Subsequent shared-risk agreements with international firms seem to have paid much closer attention to financing, and also embraced a larger circle of foreign partners in order to spread liabilities.[8]

The Tristar story illustrated an international partnership keyed to airframe and engine. Other such partnerships concentrated on the engines themselves. The first major American-European engine collaboration got under way in 1971, when General Electric and the French firm of SNECMA agreed to develop a new turbofan jet engine dubbed the CFM56. The contract called for GE to provide main engine controls and the "core engine," where fuel injection and combustion took place. SNECMA was to furnish the low-pressure engine system, the fan, gear box, and certain other accessories. Both GE and its principal American competitor, Pratt & Whitney, were strong in military engines, but Pratt & Whitney clearly dominated the commercial American market. In Europe, the commercial market was dominated by Pratt & Whitney and Rolls-Royce. Hence, GE searched for a different European collaborator. With GE, the SNECMA organization found an American firm with a respected label as well as a company that was not likely to dominate the program due to its size and market position. Still, the GE partnership was not without irony, since 10 percent of SNECMA was owned by Pratt & Whitney, the majority being controlled by the French government (80 percent) and the remainder owned by private investors.

For GE, cooperation with a European company was virtually essential. As American purchases of military aircraft dwindled, GE's existing customer base would inevitably decline. The company needed assistance to gain a foothold in the commercial market, but U.S. antitrust laws made an American partnership unlikely. Moreover, the SNECMA venture addressed another worrisome possibility. In Europe, the growing strength of the European Economic Community raised the specter of a European aerospace consortium seizing the lead in the development of engines in the 20,000- to 30,000–lb thrust class, then erecting tariff barriers to freeze out the American competition in a market estimated at

$10 billion. Accordingly, the U.S. government approved GE's request for a provisional license permitting the export of technical data.

At this point, the deal nearly fell through. GE's engine core for the CFM56 project was based on an engine (the F101) it had specifically developed for the B-1 bomber. Utilizing some $600 million in government R&D funds, the engine core symbolized the ultimate in American engine technology. On further examination, GE's provisional export license was formally suspended by the State Department's Office of Munitions Control on the grounds that it would be a breach of national security. From France, President Georges Pompidou promptly sent a protest note to President Richard Nixon. The French muttered about canceling various Franco-American projects under way, but this seems to have been more bluster than anything else. Although SNECMA pointedly sounded out Rolls-Royce and the Italian firm of Fiat about a similar engine, the French made certain that lines of communication to relevant American agencies remained open.[9]

As U.S. officials pondered the realities of a European consortium operating behind an EEC tariff barrier, second or third thoughts eventually prevailed. By altering some features of the B-1 bomber engine and adding others from a commercial GE jet engine already in production, national security concerns were alleviated. Moreover, the new agreement called for GE to reimburse the Department of Defense $20,000 for each CFM56 engine sold, addressing the issue of public tax dollars originally having contributed to GE's engine technology. In the spring of 1973, Pompidou met Nixon at a conclave in Iceland, where they formalized the GE-SNECMA arrangement. The joint venture moved rapidly ahead, with the first CFM56 engine run by GE in 1974; production units have since powered dozens of airliners around the globe. In pooling their resources, the French and American companies developed an engine that neither would have pursued alone. Both were able to enlarge their presence in the commercial market, and the United States established a significant beachhead inside the EEC. America's willingness to pursue the program after initial collapse underscored growing awareness of the significance of international markets and international joint ventures.

Balancing concerns about foreign components overwhelming the U.S. market, there was the reality of American engines and avionics having had a major role in Japanese, Canadian, Latin American, and European aircraft for years. Moreover, this kind of subcontracting had been growing increasingly widespread and diverse. The new Fokker 100, a Dutch-built airliner of 107 seats, incorporated products from Goodyear (wheels and carbon brakes), Grumman (engine nacelles and thrust reversers), Garrett (auxiliary power units and air conditioning), Sundstrand (electrical supply system), and Teledyne (flight data acquisition unit). Underscoring its international complexion were the risk-sharing European contractors like Shorts (Irish manufacturers of the wing), Rolls-Royce (the engine supplier from Britain), and MBB (the German manufacturer of major portions of the fuselage and the entire tail unit, except for the rudder).[10]

The Airbus Industrie family of air transports did not originally incorporate major American-made structures, although thousands of U.S.-made components became part of each airplane. In 1984, when Pan Am announced orders and options for 91 Airbus transports, there was strong reaction that the American airline was not being patriotic. Airbus Industrie immediately took double-page newspaper ads across the United States, reminding Pan Am's future customers that the Airbus possessed a significant Yankee connection. The ads stressed that 500 American companies in 35 states had products on Airbus planes, including options for indigenous American-made GE engines or the international CFM56 units.[11]

When the McDonnell Douglas MD-11 tri-jet entered service in the early 1990s, the international character of American airliners was further demonstrated. The manufacturer offered a choice of American or British engines. Fuselage skin panels, from nose to tail, were fabricated by Aeritalia, who also built the winglets and the vertical stabilizer; CASA of Spain supplied the horizontal stabilizer, along with main landing gear doors. Japanese firms fabricated the tail cone and outboard ailerons; outboard flaps came from Brazil; and British firms supplied additional flap components, as well as the nose gear. McDonnell Douglas of Canada built major wing assemblies and wing box fuel tanks. From the outside, a casual observer of the MD-11 would be able to spot only a few surfaces that were originally "made in the U.S.A." Moreover, many of the "subcontractors" were also risk-sharing partners, having put up considerable cash or contributed production capacity.[12]

In retrospect, internationalization expanded far beyond the original practice of subcontracting. Given rising production expenses, American firms willingly broadened their horizons to include risk-sharing with foreign manufacturers of structural subassemblies, and, in the case of Lockheed, an entire program on a careful integration of airframe and engines. The GE-SNECMA collaboration was pursued to give the United States better market penetration abroad, and this increasingly became the case for aircraft manufacturers like Boeing and McDonnell Douglas. With approximately two-thirds of the market for commercial transports outside the United States, the Aerospace Industries Association was very clear about the future of international collaboration, stating the "U.S. aerospace companies . . . must participate in cooperative relationships to maintain a healthy, competitive industry."[13] This symbolized a historic shift in the pattern of American aerospace manufacturing of commercial airliners.

Space and NASA in the Post-Challenger Years

As NASA entered the post-*Challenger* era, it did so under a leadership that had changed in demographics and character during the late 1980s. NASA had expected the departure of many skilled personnel who had committed to stay only until the shuttle resumed flights, but other factors came into play, as well. Veto of a federal pay raise meant problems in recruiting many qualified people, and new ethics rules preventing high-level managers from working for certain contractors for up to two years after leaving the agency also threatened the recruitment of key executives. A hiring freeze in the mid-1970s led to a shortage of managers in the 30- to 35-year-old age group, and so, during the years 1989 to 1991, when about 70 percent of the agency's senior and middle management was due to retire, a serious gap in agency leadership existed. Some observers saw the possibility of restructuring the agency by making field centers into contractor-operated facilities in the future. These and other issues fell to subsequent NASA administrators.[14]

Meanwhile, activities in space involved continuing programs from NASA and military sources, although commercial operations made remarkable progress, beginning with communications satellites. At the end of World War II, the British scientist and writer Arthur C. Clarke made a realistic proposal for communications satellites. Based on the possibilities represented by the V-2 rocket and other trends, his article in a 1945 issue of the journal *Wireless World* suggested satellites functioning in stationary, or geosynchronous, orbit above the earth. Following Sputnik I in 1957 and subsequent events, American Telephone and Telegraph paid for development of the 170-pound Telstar 1, a trial communications satellite launched by NASA on July 10, 1962; Telstar 2 followed a year later. These and several successors (Syncom and Relay) represented technology demonstration units, with short lives in low orbits, although their voice and image transmissions proved exciting. In 1964, the Syncom 3 satellite delivered a live broadcast of the opening ceremonies of the Olympics in Japan. That same year, the United States helped organize the International Telecommunications Satellite Organization, owned by member states. The U.S. organization, a public-private enterprise called Communications Satellite Corporation, or COMSAT, became the group's operational arm, with headquarters in Washington, D.C. Over the next few years, a series of Intelsat units were carried into orbit to begin regular international operations. Early versions possessed 240 circuits or one television channel, with monthly costs of $30,000 per circuit. By 1977, Intelsat had invested more than $437 million and sponsored some 27 launches; 12 satellites remained in use by 102 members; the utilization cost per half circuit came to only $570 per month.

During the 1980s, the Federal Communications Commission granted considerable latitude to commercial development in the United States, so that customized satellites were built to accommodate shipping lines, commercial communications networks, and sundry businesses who needed to transmit high volumes of communications traffic over several hundreds of miles. Domestic satellites included Westar (Western Union), Comstar (AT&T and COMSAT), and Satcom (RCA and partners). Newspaper wire services converted their operations to satellite transmissions, and the eventual growth of cable television networks exploded during the 1980s thanks to direct satellite distribution networks. Comstar 1, which weighed more than 3,000 lbs, could handle up to 6,000 telephone calls and 12 television programs in various combinations simultaneously. With Hughes Aircraft Company as prime contractor, the unit was operated by Comsat General and leased by American Telephone and Telegraph Company, which marketed its circuits to a variety of customers. Hughes, Lockheed, Martin Marietta, General Electric, and other contractors met the needs of a sizable national security market as the 1960s through the 1990s brought orders for a panoply of military communications and early warning satellites, in addition to proliferating reconnaissance or "spy satellites."[15]

For some years, the civil payloads arrived in orbit aboard NASA boosters, including the space shuttle. In the late 1980s, launches of commercially marketed boosters became considerably more active, with vigorous competition offered by European organizations such as ESA and its Ariane booster. Following the *Challenger* explosion in 1986, the Reagan administration initiated new policies to take NASA out of the business of rocket-launched commercial payloads, relying instead on contracted services with private corporations. The final pair of NASA unmanned rockets blasted off in the autumn of 1989—the last of more than 400 NASA rocket missions from Cape Canaveral.

Meanwhile, the first commercial American rocket launch had already taken place at the Cape. On August 8, 1989, the McDonnell Douglas Astronautics Corporation, using one of its own reliable Delta boosters and its own crew, launched a $150 million British communications satellite into geostationary orbit above the mid-Atlantic. Although several new organizations proposed launch services, aerospace heavyweights like McDonnell Douglas, Martin Marietta, and General Dy-

Launch of a Titan IV intercontinental ballistic missile. Originally delivered by Martin Marietta (now Lockheed Martin) as the prime contractor, the Titan family served as the launch vehicle for nuclear warheads, military satellites, NASA astronauts and space research missions, and, in the 1990s, commercial payloads for American as well as foreign customers. *Courtesy of Lockheed Martin.*

namics became the first to put payloads into orbit. Martin Marietta's turn came early New Year's Eve 1989. The company's Titan 3 booster carried a dual payload: one satellite from the British Ministry of Defense, and a second from a Japanese organization, to be used for telephone, television, facsimile, and high-speed data services. These and other projects in the works appeared to send commercial space ventures off to a robust start. Shuttles could stay in orbit, conduct manned experiments, facilitate repair of spacecraft, and retrieve payloads, but for straightforward payload-to-orbit deliveries using a single, expendable booster, vehicles like the Titan IV cost only half as much as a $500 million shuttle mission.[16]

International agreements snowballed. Despite political protest in the aftermath of the Chinese government's suppression of student dissidents at Tiananmen Square in June 1989, President Bush lifted a ban imposed on the launch of three communications satellites by a Chinese booster. Manufactured by Hughes Aircraft, two were for an Australian company and one for a consortium in Hong Kong; all were orbited in 1990–92. During the early 1990s, commercialization of former Soviet national programs accelerated a startling variety of Russian-American projects. Lockheed Missiles and Space Company negotiated a joint venture with a recently privatized Russian firm, Energia, to utilize the latter's Proton booster for launches of various commercial payloads. In short order, the venture lined up nine firm launch orders worth $650 million during the late 1990s. In a different context, other U.S. companies urgently searched for means to upgrade American capability for expendable launch vehicles (ELVs). Talks got under way with other Russian organizations about employing their brawny, liquid-fueled engines for the first stage of new ELVs using upper stages supplied by Martin Marietta, and also as power for an upgraded single-stage Atlas built by General Dynamics. This irony—the Atlas representing America's ICBM deterrent to the Soviet threat of the 1950s—exemplified the yeasty atmosphere of the post–Cold War era.[17]

Space science continued its forward march, with NASA planning several dozen major science missions to Venus, Jupiter, and elsewhere during the late 1980s and early 1990s. Foreign partnerships and contributions played a key role in the planning and conduct of these programs. The European Space Agency (ESA), for instance, supplied funding, equipment, and personnel for the Ulysses project to map Venus.[18]

In the course of these ambitious undertakings, NASA—and the American space program—ran into serious problems. One of NASA's most publicized projects, the Hubble Space Telescope, was plagued by cost overruns and delays. Finally orbited in 1990, flawed components threatened to make it useless until a last-ditch repair mission by a team of astronauts salvaged the project. With vastly improved images as a result, NASA breathed more easily, but its reputation took a beating in the scientific community, the press, and congressional budget debates. Meanwhile, NASA's biggest program, the manned space station, began to come under serious scrutiny. Given the long lead time expected for the station's development, NASA was compelled to accede to congressional and corporate initiatives for many project modifications along the way, in order to sustain popular support and appropriations. The dictates of "incremental politics" meant considerable frustration for scientists and engineers, who grappled with changing blueprints, schedules, and costs. President Ronald Reagan gave budgetary approval to an ambitious space station in 1984; the high costs of research and development motivated NASA to solicit foreign partners, leading to a prolonged series of prickly negotiations with Canada, ESA, and Japan. By

1990, space station *Freedom* had consumed $8 billion in R&D contracts (with a projected cost of $30 billion) and remained in planning even though operations had been scheduled for 1992. The *Challenger* loss, Hubble's woes, and a rash of other projects with missed schedules and cost overruns, such as a geodetic satellite at 230 percent over budget, all put NASA on the defensive. Late in 1990, an adamant Congress ordered NASA to revise its plans and develop a less complex and less expensive station.

While NASA backpedaled in the face of congressional pressure, the agency also became the target of an increasingly hostile press. The fact that the space program was no longer perceived as a crucial national interest in counterpoint to Soviet initiatives undoubtedly contributed to a decrease in its popularity. Even the name for the American station, *Freedom*, seemed needlessly chauvinistic by the early 1990s. Nonetheless, much of the ire directed at the agency came from a lingering feeling that manned space missions had consumed far too much of NASA's budget and its attention. In 1980, NASA's budget amounted to $5 billion, rising to $10.7 billion in fiscal year 1989. Manned space efforts absorbed more than half of the 1989 budget, with space science allocated $1.9 billion, or about 18 percent; the figures reflected a familiar pattern over the years, with space science receiving about 20 cents of each NASA dollar. Now, both the manned enterprises as well as space science agendas seemed to be in alarming disarray.[19]

Elected in 1992, the new Clinton administration added criticism from both the White House and the Office of Management and Budget. A new NASA administrator, Daniel Goldin, took over, coming from a background of extensive corporate experience with aerospace contractors like TRW. During the spring of 1993, Goldin launched aggressive plans to devise a new blueprint for NASA's future and to re-organize the space station venture. A high-level space station committee soon scrapped plans for a permanently manned station in favor of a far more modest structure to be periodically occupied in the pursuit of extended research. Significantly, Russian participation emerged as a strong possibility, and the *Freedom* was renamed *Alpha*. Goldin revamped NASA's bureaucracy as well as space station management, and announced Boeing as the manufacturer that would manage all other space station contractors, streamlining a heretofore multilayered and massively complex process. NASA's prior system of management, averred Goldin, had been "appalling."

Further, NASA turned resolutely to trim its budget, slash a list of space-related activities, and divert more of its money toward aeronautics. Aviation had come more to the forefront in the wake of solid European gains in the global marketplace for civil aircraft. In 1969, American builders delivered 91 percent of the world's civil transports; the U.S. share in 1993 had dropped to 67 percent. Representatives of industry pointedly argued that the aerospace business employed more than one million workers and sold more than $90 billion worth of aircraft and related products. As military contracts continued to shrink, the need to maintain American competitiveness in the civil sector got congressional attention.[20]

The aeronautical initiative already possessed a respectable momentum; NASA's flight research had never become moribund. In cooperation with the Department of Defense, NASA continued to study a winged, hydrogen-fueled aircraft capable of cruising above the earth at Mach 12, streaking into earth orbit at Mach 25, and, in a passenger transport version, speeding from America to Asia in three hours. Dubbed the "Orient Express" by the news media, this National Aero-Space Plane, or NASP, remained at the R&D level from the 1980s through the 1990s. A variety of other projects were aimed at more immediate applications. During the same time period, defense-related programs included fly-by-wire systems, high-speed aerodynamics, exotic wing planforms, and thrust control studies for combat planes in cooperation with other countries like Britain and Germany. Special teams continued to study rotary aircraft, with particular attention paid to the innovative V-22 tilt-rotor twin-engine designs flown as full-scale prototypes, and built by the engineering team of Bell Helicopter and Boeing Vertol.

NASA also sustained its engine research with an eye to more economical functioning, less noise, and reduced exhaust. Studies of the icing phenomenon persisted, and different programs yielded significant insights into issues of cockpit workload and design. In a program aimed specifically at addressing commercial airline needs, NASA acquired a Boeing 737 and rigged a special test cockpit in the plane's passenger cabin. While a safety crew rode in the original cockpit as a backup, pilots in the experimental cockpit module worked with various advanced systems for flying transports in high-density airport environments; approach and landing procedures in bad weather; and other operations in the increasingly crowded airspace of national and global routes. A long roster of additional activities continued, sharpened by the heightened sense of competition from overseas.[21]

The Defense Business and Desert Storm

Regrettably, reports of corrupt practices continued to haunt the aerospace industry, still under a cloud due to dramatic news stories about cost overruns and the government's bailout of Lockheed. Questions of illegal activities swirled around Northrop's 1980s efforts to sell its F-20 Tigershark fighter, a program in which the company had invested $1.2 billion of its own funds and which desperately needed sales. Northrop's earlier multimillion dollar efforts to sell the F-5 had contributed to the clamor for regulation in the 1970s.

Clearly, Northrop was not alone. During the mid-1970s, investigations related to the Watergate scandals uncovered extensive overseas bribery by the Lockheed Corporation, totaling some $38 million. Eventually, McDonnell Douglas and staid Boeing were implicated in the paying out of tens of millions of dollars to agents overseas to grease the way for foreign sales. It made little difference that other industries did the same or that payoffs and kickbacks were imbedded in the business practices of many international cultures; the high-profile aerospace industry became the target of exposés in newspapers and popular magazines. The furor over these events led to the Foreign Corrupt Practices Act, enacted by Congress in 1977 to halt bribery overseas by U.S. firms.

But corporate skullduggery persisted, and Northrop Corporation continued to appear in the sights of various prosecutors. The company's sales came to $263 million in 1959, rising to $5.8 billion in 1988. During the Reagan years of defense building in the 1980s, Northrop garnered some $36 billion worth of contracts. Unfortunately, Northrop and other contractors did not always act responsibly. During the late 1980s and early 1990s, a federal grand jury conducted wide-ranging probes into illegal practices in the aerospace industry, an endeavor called "Ill Wind." The investigations revealed illegal activities by such firms as Hazeltine Corporation and Teledyne Electronics, which pleaded guilty to resorting to bribery while seeking inside information about U.S. Navy electronics contracts. Early in 1990, Northrop admitted guilt to 34 felony counts involving falsified test reports relating to air-launched cruise missiles and AV-8B Harrier jets.

Against such a background, the defense industry's huge budgets made aerospace contracts highly controversial in the press and in Congress, especially since many seemed to have soaked up hundreds of millions of tax dollars over an inordinately long period of development. American military operations in Grenada (1983) and Panama (1989) seemed awkward, with lackluster performance by high cost, "high-tech" combat equipment. Stung, the military services and their contractors modified their procedures, creating an improved fighting force with an arsenal that turned in a convincing performance when America became involved in the Persian Gulf conflict.[22]

During the summer of 1990, Saddam Hussein of Iraq launched an invasion across Iraqi borders into the neighboring oil fields of Kuwait.

Under United Nations auspices, the United States and coalition allies shipped garrison forces into the region. When Iraqi troops ignored U.N. deadlines to evacuate Kuwait, coalition military units unleashed Operation Desert Storm to repel Iraq's aggression, liberate Kuwait, and stabilize the region. Desert Storm, launched in early 1991, triggered a variety of responses in the American public. There were many who opposed the military action because of the inevitable destruction and civilian casualties, as well as the military losses. On the other hand, Desert Storm represented an awesome display of high-tech weaponry that mesmerized press and public alike. Before the war, there had been much criticism of Reagan's skyrocketing defense budgets, to say nothing of the Pentagon's arrogance. There were irritating stories of consistent defense "contract overruns" in money for equipment that reportedly had serious flaws. But Desert Storm's weapons enjoyed unprecedented media coverage—sometimes live from the battlefield via satellite—and the weapons seemed awesome.

Iraq's armed forces proved to be less formidable than supposed, and some of the advanced weaponry deployed by allies like Britain and France, in particular, was most impressive. Still, the U.S. military effort held center stage and seemed to capture the nation's fancy. As *Fortune* magazine shrewdly noted, "America has discovered its arsenal—the damnedest array of stealthy, micro-processed, laser-guided, thermal-imaged, satellite-vectored weaponry ever imagined." Much of the enthusiasm for this new arsenal seemed to stem from the realization that American technology was not only indigenous but obviously the best in the world. *Fortune* went on to explain that Americans celebrated the fact that we were "still capable of manufacturing something better than anyone else—even if we can't drive it off or play a videotape on it."

The defense industry watched all of this with satisfaction. For one thing, its products looked good, the Iraqis were quickly routed, American casualties were low, and the defense industry suddenly enjoyed good press. In the short term, a good deal of the equipment and munitions would need replacement, thereby keeping production lines open. In the long run, noted some hawkish observers, the nuclear threat might diminish, but many Third World countries would be keen to equip themselves with the advanced conventional hardware demonstrated in the Desert Storm engagement. As Third World forces modernized, it behooved the United States to maintain a technical edge if yet another showdown occurred 10 or 20 years in the future. Thus, despite anticipated declines in many areas of the defense industry, including aerospace, a considerable share of the "peace dividend" was expected to disappear into new military spending.[23]

For all their appeal, none of the Desert Storm weapons were cheap, and many had experienced long and frustrating development. A good deal of the Pentagon's weaponry originated in the Carter administration. The argument in those days followed the line that high-quality U.S. equipment would compensate for the quantitative edge enjoyed by the Soviet Union. Despite the inherent high costs, quality became the watchword in new weapons systems. During the Reagan years, quantity as well as quality was pursued, and defense budgets accelerated in meteoric fashion. Procurement contracts seemed to offer innumerable opportunities for corporate mischief, so that press and public alike came to view weapons programs as inordinately time-consuming and grossly expensive. Consequently, the Pentagon added layers of oversight bureaucrats, which, ironically, contributed to yet slower progress, more stacks of paperwork, and further ballooning of costs. For the Amraam missile (advanced, medium-range air-to-air missile), initial specifications ran to a numbing 500,000 pages. Even more paperwork, and several billion dollars, had been expended on the complex design of the A-12 Stealth aircraft, under development by McDonnell Douglas and General Dynamics. Proposed for naval carrier operations as well as Air Force strike missions, the A-12 seemed continually enmeshed in shifting technological and operational requirements. Although Desert Storm helped the Pentagon justify several criticized programs, President George Bush's new Secretary of Defense, Dick Cheney, decisively changed acquisition procedures. He canceled the controversial A-12 and supported new procurement arrangements based on recommendations from a high-level panel (the Packard Commission) for more realistic specifications and prototype hardware.

As the cardinal weapons merchant to the Department of Defense, McDonnell Douglas was able to tout an unusual variety of weapons during the Gulf War. From its corporate offices in St. Louis, the company operated more than 50 plants scattered from Long Beach to Toronto. In 1989, the company's revenues came to $14.6 billion, of which $9 billion, or 60 percent, came from weapons. McDonnell's military staples included five different missile types, seven different airplanes (combat and transport) and a laundry list of miscellaneous military gear. During the Gulf War, the omnipresence of its products led some punsters to label the U.S. assault "the big Mac attack." Skeptics focused considerable attention on McDonnell's AH-64 Apache attack helicopter, produced at $10 million each. Earlier, the Government Accounting Office reported that the complex chopper would be too unreliable to operate effectively in sustained combat. The AH-64, which had been plagued by development problems, was one of the many high-priced, high-tech aerospace weapons that could make or break the quality theory of weapons acquisition.

Developed originally by Hughes Helicopters, the Apache emerged in the mid-1970s as a consensus design using advice from Vietnam chopper pilots and advanced planning teams looking ahead to the need for capable antitank weapons. High on everyone's list was the need for night-fighting capability—a special problem, given the chopper's normal operational arena close to the ground, where innumerable trees and other obstacles lurked. Intense research and development perfected an infrared targeting and night vision system that allowed the pilot and the gunner-navigator to fly safely at night, as well as in dense fog, rain, and falling snow. The targeting equipment permitted the Apache's two-man crew to pick out enemy targets dozens of miles away; laser-guided Hellfire antitank missiles virtually guaranteed a hit.

The Apache production process was not a smooth one. Airframes came from a Teledyne plant near San Diego; engines from General Electric in the Midwest; avionics and fire control from Martin Marietta, Northrop, Honeywell, and others. Hughes itself built the rotors and related dynamic components, and the whole aircraft came together at a plant in Mesa, Arizona. Its awesome electronic gear took considerable time to install and test, to say nothing of the engines, flight instruments, and extensive armor for crew protection. McDonnell Douglas, interested in market prospects for both civil and military helicopters, bought out the Hughes helicopter division in 1984, paying about $480 million. Despite McDonnell's experience in military programs, the process of completing AH-64 development presented a steep learning curve, and McDonnell Douglas announced a $107 million write-off in 1989 that represented cost overruns on the Apache effort.

In the spring of 1990, Thomas Gunn, described as a businessman-lawyer, was named president of the new division and received instructions to put things in order. Most of the workers had been at the Mesa plant since its opening in 1983. They were notably young, with an average age of 28, but they finally had begun to mature. For all that, stringent economics were required, and that meant trimming personnel from top to bottom. Gunn fired 1,400 of 7,400 employees, many from management, and reorganized plant procedures. After surveying several possibilities, Gunn's management team settled on a production practice in use at Volvo automobile plants in Sweden. The Volvo plan centered on teams that built a complete car, rather than relying on an assembly-line basis. Workers at McDonnell Douglas quickly became converts to the team approach, citing the chance to learn more skills in the manufacturing process and parlay those skills into higher wages. If a team member became sick or took vacation leave, other team members with equivalent skills filled in, so that production remained on schedule. Morale at the Apache plant shot up and the company began reporting firm profits again during 1990.²⁴ The Apache became a valued asset during Desert Storm.

At the same time, persistent antagonisms among various nations around the globe suggested continuing sales potential in the export market. For Raytheon, another major defense contractor, the future looked profitable in the Gulf War's aftermath because of the spectacular reputation of its Patriot antimissile system. As global television broadcast the ominous wail of Israeli air-raid sirens, television screens recorded the brilliant flash of Patriots in high-altitude encounters with Iraq's notorious Scud ballistic missiles. Although some Scud attacks created damage and military casualties, the Patriot's apparent ability to obliterate incoming missiles in midair made it a media success and a highly marketable commodity. As one senior analyst from the Brookings Institution remarked, "I suspect there will be other countries around the world with concern about their neighbors having missile technology. They're going to want these things." Raytheon's annual income totaled $9 billion, with 55 percent coming from the military marketplace. With Patriot systems ringing up some $1.5 billion in 1990, before Desert Storm, the company's financial outlook was bullish. Other advanced weapons, like the Lockheed F-117A Stealth attack fighter ($42.6 million per plane), KH-11 spy satellites, and a catalog of U.S. Air Force electronic systems, also operated convincingly.²⁵

Nevertheless, the extraordinarily high research, development, and procurement costs of American weaponry had an impact on the next generation of combat planes to be acquired by the Air Force: the Northrop B-2 Stealth bomber and the Advanced Tactical Fighter. The B-2 represented the "paper airplane competition" process that had fallen in and out of favor since World War II. Rather than build prototype bombers to fly in competition, a prohibitively expensive approach as large jet aircraft became more complex and costly, the armed services turned towards competitive proposals instead. In this situation, military planners drafted an elaborate document that specified formal requirements. Competing contractors submitted written proposals that ostensibly described how they were going to meet—or optimistically exceed—these requirements. A prime manufacturer ultimately received a sole contract to build and produce a plane that had yet to get off the ground. It was a system that led to unfounded promises, all sorts of production snafus, and spiraling costs. The Navy A-12 Stealth attack plane was one of the casualties of this development process. Before the A-12 was finally canceled in 1991, the Navy had spent some $5 billion and was left with little more than shelves of engineering studies and stacks of computer printouts. Although the B-2 got into the air and boasted a number of technological innovations, its estimated price tag of $865 million per plane triggered a congressional revolt. Late in 1991, Congress recommended that the 15 bombers already under production would constitute the entire fleet, rather than the 132 aircraft originally planned. To many observers, the B-2 development process represented the end of an era, especially when compared to the dazzling success of the process followed in the development of the Advanced Tactical Fighter.

The Advanced Tactical Fighter (ATF) program was just getting under way in 1986 when the presidential commission directed by industrialist David Packard issued its striking report on weapons procurement procedures. The Packard Commission judiciously criticized the "paper airplane" approach, especially when a plane represented a technological leap, but produced no prototypes to demonstrate that such a leap really worked. The commission also criticized overly detailed specifications because they often discouraged new approaches that might save time and money. One recommendation was included that to many in the defense industry seemed downright deviant—the audacious notion that contractors themselves should cover a good share of the R&D expense in constructing prototypes. The commission reasoned logically that contractors would be less inclined to overspend and more inclined to introduce cost-effective procedures.

Despite resistance within the Pentagon and from contractors, the defense industry was already in such ill repute that pressure to implement Packard Commission ideas forced some sort of gesture. The ATF program became the testing ground. Detractors of the Packard Commission inwardly smiled; successful flight of a prototype ATF with supersonic speed and high maneuverability seemed problematic at best, given the nagging problems of the stately subsonic B-2 Stealth bomber. But there were others in the defense establishment who wanted to take the Packard Commission's report and make its recommendations work. Within four years, the ATF prototypes (Lockheed YF-22 and Northrop YF-23) became star performers as they careened through supersonic aerobatics in the skies above Edwards Air Force Base. Lockheed and Northrop managers had both realized that they had to put up considerable funds for the ATF program, and that the loser absolutely would not get any compensation. In the end, the YF-22 and YF-23 projects each got about $818 million from the government but had to invest roughly $1 billion of their own and their partners' money. The high costs led to interesting teaming agreements, with Lockheed's team including General Dynamics and Boeing; the Northrop design represented a partnership with McDonnell Douglas. In addition, engine manufacturer General Electric squared off against Pratt & Whitney in the power plant competition, and they each had to ante up private capital as well. But the winners could hope for $93 billion in contracts for 640 planes.

In the final analysis, the Lockheed YF-22 fitted with Pratt & Whitney's potent new engine emerged the winner. Because both planes and both new engines performed so well, and because the whole process took only four years with a relatively low cost of $3.9 billion, the ATF program inevitably invited comparison with the contentious B-2 exercise. In other words, was the Packard Commission on target with its criticisms and recommendations? From project start to first flight, the B-2 bomber consumed eight years and $33.2 billion. There was no competing prototype—no needling rival to keep Northrop (as sole source contractor) looking over its shoulder. There were offsetting factors. The ATF designs enjoyed a legacy of Stealth technology from the B-2; the bomber was admittedly larger in structure and its flying wing configuration presented some special control problems. Also, the B-2 expense totals included production tooling. As the F-22 entered manufacturing, the total ATF bill could expect to escalate by several billion dollars. Cynics remarked that both the Air Force and the contractors underplayed the ATF success because it made the sole-source B-2 effort look so bad. Still, the ATF's reconfigured management, plus the undeniably impressive record of fielding two advanced prototype planes and two different engines in record time with controlled costs, seemed bound to have an impact on future acquisition programs.[26]

The manufacturing and assembly of airliners—engines and electronics as well as airframes—involved an intricate network of partners, suppliers, and production agreements encircling the globe. Foreign companies turned out an impressive array of components for American planes; American companies supplied a variety of major aircraft programs overseas. Many of these joint ventures commanded so much capital investment and national prestige that when problems arose, they were ironed out by the political leadership of the governments involved.

As for America's space program, NASA continued to experience political problems while recovering from the *Challenger* tragedy. The 1990s brought comprehensive management changes and redefined goals, including a new emphasis on aeronautics to help the United States meet the challenges of global competition. In the meantime, privatization of launch services closed a dramatic chapter in NASA operations, while a startling variety of international agreements underscored the new cooperation between America and the former Soviet Union. The increasing market for commercial satellites symbolized the day-to-day influence of astronautics on society.

The Gulf War became a showcase for an impressive array of new weapons like the AH-64 helicopter, F-117A Stealth aircraft, and a variety of wizard-like electronic and computerized combat systems. Behind it all lay new concepts of production management adopted by American firms. Additionally, stringent budgetary guidelines required new approaches in meeting military procurement policies. Theoretically, all this helped the civil and defense components reach a state of preparedness for the twenty-first century. At the same time, the structure of the aerospace industry in America and overseas experienced immense change, and its appearance in the twenty-first century would be much different.

CHAPTER 9
Twenty-First-Century Transitions

WITH AEROSPACE AT the threshold of its second century of activity, the future seemed unusually difficult to see in clear focus. The political turbulence of the 1990s was reflected in virtually every aspect of aerospace activity, from manufacturing to air travel. On the one hand, continued U.S. and Soviet deescalation forced extensive layoffs and cutbacks throughout the defense industry. On the other, costs of high-tech aerospace weaponry, export sales to Gulf countries, expanding markets throughout Asia, and continued growth of air travel requiring new transports, were expected to sustain American aerospace industry income in the future, although at notably lower rates.

Aerospace manufacturing in the United States recorded sales of $140 billion in 1991, with commercial sales exceeding defense sales for the first time since 1980. With a backlog of $235 billion, the industry accounted for about half of all unfilled orders for U.S. durable goods in 1991, although analysts projected lower aerospace sales in real terms during the 1990s as compared to the 1980s. The global aerospace industry faced massive changes in a drastically changed business environment. The fragmentation of the former Soviet Union and the end of the Cold War triggered deep cuts in defense spending that continued to shake manufacturers throughout the former Communist bloc nations, western Europe, and the United States. Moreover, a sluggish global economy caused problems for civil aviation operations, as well. As a result, many airlines canceled orders for new planes and stretched out delivery dates of those on order, adding to the difficulties for aerospace manufacturers in Europe as well as in America. Analysts predicted that the next several years were bound to include mergers, consolidation, and serious restructuring within the surviving companies. Fierce competition for a shrinking aerospace market was also expected to trigger even more international joint ventures as a means of spreading risk and gaining a foothold in foreign markets.[1]

The International Marketplace

Foreign sales appeared to offer a growing market for American helicopter manufacturers, even though the competition from overseas companies remained intense. By the mid-1990s, Bell Helicopter reported 42 percent of its deliveries to the U.S. military services and 58 percent in commercial sales, of which 78 percent came from outside the American market. Sikorsky's sales in 1993 also showed strength overseas, rising from 5 percent to 30 percent over a five-year period. Both companies had drastically downsized due to the decline in military orders, and looked to commercial buyers and exports in order to sustain production. Kaman Aerospace (helicopters) stayed in business as a component manufacturer and as a builder of specialized machines for external lift operations such as logging and construction.

McDonnell Douglas Helicopter Company continued production of the AH-64 Apache and concurrently exploited its unique Notar anti-torque system of control on smaller commercial and military models. "Notar" (a contraction of "no tail rotor") became a major marketing tool in the company's commercial sales efforts, in particular. The Notar system replaced the conventional tail rotor with what McDonnell literature described as a "fan-driven air-circulation system within an enclosed tail boom." The effect was to create a strong air flow out of the tail that interacted with the main rotor's downwash to create a control effect. The Notar system reduced the danger of tail rotors to ground crews, facilitated low level operations where a tail rotor might collide with obstructions, and reduced external noise by as much as 50 percent. The last factor became a central point in sales literature. In 1991, the McDonnell MD 520 series of single engine, five-place helicopters introduced the Notar system to customers, beginning with the delivery of the first of seven units to the Phoenix Police Department. A twin-engine, light-place model began reaching customers in 1994, with strong sales to emergency medical service operators.

Other manufacturers worked on improved efficiency, especially techniques to reduce noise. It seemed likely that more would also internationalize their manufacturing plans. In the case of McDonnell Douglas, the new twin-engine MD Explorer incorporated components supplied by risk-sharing partners around the world. One organizational scheme listed the McDonnell Douglas Helicopter as responsible for design integration, assembly, and delivery. Component production occurred elsewhere. There was only one American partner: Pratt & Whitney, supplier of the engines. An Australian firm built the fuselage; transmission and other machinery originated in Japan; interiors in Britain; instrument displays in Canada; and other key components from partners in Israel and France, as well as additional firms in Japan, France, and Canada, for a total of 10 foreign partners in all.[2]

The story of Piper Aircraft, once a leader in the general aviation industry, underscored the diverse pitfalls in the aerospace manufacturing business. Piper filed for Chapter 11 protection early in July 1991, even though it had assets of $70 million, liabilities of $30 million, and an order backlog of some $170 million. As orders rolled in, the cost of raw materials also rose, and Piper's credit lines with suppliers eventually ran out. Several companies abroad expressed interest in buying the company, so long respected in the history of American aviation. Negotiations with U.S. and foreign investors like Aerospatiale came to nothing; Piper reported that potential suitors were frightened off by the rising costs of the liability suits that had plagued manufacturers over the past decade. With only 200 employees left out of a workforce that once numbered 3,200, Piper's owner continued to seek an investor and an opportunity to cash in on existing orders for hundreds of aircraft. A special court order permitted limited production to resume during 1992, keeping Piper's struggle alive. In the meantime, Gulfstream Aerospace, Beechcraft, and Cessna all continued to stress production of high-value corporate turboprops and jets. Cessna pushed ahead with design of the Citation 10, a large corporate jet with a cruise speed of Mach 0.9, just under the speed of sound, and tentatively priced at $12.5 million. Learjet, under the Canadian firm of Bombardier, enjoyed a new lease on life in the mid-1990s with arrangements to deliver a dozen aircraft for a Navy training program; it also announced an encouraging production backlog of some $350 million. Gulfstream's executive jets in the $20 million category continued to find buyers.[3] Beechcraft's corporate parent, Raytheon, struck a new deal with a former overseas partner to expand its line of executive jets. British Aerospace had made significant changes to its Hawker 125 model, and in 1993 Raytheon took over final completion and distribution of the new plane, called the Hawker 1000. Priced to sell at about

$13 million, the luxury jet was the first to be marketed under the umbrella of a new division, Raytheon Corporate Jets.

One new market opening appeared in the form of special-purpose aircraft, in which the buyer (usually a government) bought corporate jets "off-the-shelf" and added various types of cameras, sensors, and electronic gear for various purposes such as maritime patrol. Over several years, Learjet delivered more than 200 such planes to be fitted later for high altitude photography, target towing, combat simulations, electronic warfare training, and other tasks. The industry also explored multiple leases on an individual plane, called fractional sales, which allowed corporations to boast about "our own executive jet" while holding the line on costs. These sorts of ventures, coupled with several beneficial pieces of legislation, offered a gleam of hope in the face of major disasters in some product lines and continuing discouragement about each year's production totals.

The distinctive Beech Starship, greeted with such enthusiasm in the early 1990s, failed to win corporate success. Beech had turned out 47 Starships by the autumn of 1993, but sold only 16; the company stopped production and wrote off nearly $1 billion in cost. At $3 million to $4 million apiece, the turboprop Starship remained overpriced in the market segment where Beech had planned to phase out its King Air corporate twins. To make matters worse, the avant-garde Starship also appeared in a declining market. Overall, the deliveries for the light plane industry dropped below 1,000 to 899 in 1992, another historic low, and were estimated at only 850 in 1993. Piston-engined deliveries remained flat, and few turboprops or corporate jets had posted even slight gains.

The biggest—and best—news came from the U.S. Congress, which finally enacted a favorable law on the issue of product liability. The legislation instituted a "statute of repose," a period of no more than 18 years during which the manufacturer would be liable for a product. The law finally gave credence to the industry's argument that they should not be held responsible for aircraft used beyond a reasonable product life, a fact that cleared the way for dramatic reductions in insurance liability costs. Moreover, the 1994 law covered existing piston-engined aircraft fleets in the United States, which had an average age of 27 years. Cessna announced immediate plans to reopen its production lines for some single-engine piston-powered planes, and other companies prepared to renew or increase their single-engine airplane production. While not expecting the record years of the 1970s, the light plane industry was clearly in a more optimistic mood.[4]

In Europe, where nationalized industries and extensive state subsidies had been the rule, consolidation and privatization became more extensive as governments struggled under severe budget pressures and voters were no longer willing to shoulder growing deficits in state enterprises. The new environment prompted a fresh wave of international arrangements, as foreign companies made every effort to secure a share of dwindling sales opportunities. Even in slow economic times, the North American markets remained huge and a basic degree of activity persisted. Alenia, the Italian conglomerate, branched out from its European base and acquired Dee Howard, a successful modification and maintenance company located in San Antonio, Texas, as well as other holdings. Similarly, the British firm Lucas Aerospace diversified its manufacturing capabilities throughout Europe and North America, acquiring several electronics divisions located in the United States. The French firm of Matra bought out Fairchild Space and Defense Corporation; several dozen American firms went to European conglomerates by the mid-1990s.

The competition for remaining aerospace defense contracts remained particularly keen; as in previous decades, U.S. companies hoped the use of a major contractor abroad would stimulate additional foreign sales. The U.S. Air Force and U.S. Navy solicited proposals for the Joint Primary Aircraft Training System (JPATS), which would carry with it a very large order for an estimated 700 or more planes. In an industry increasingly starved for orders, the JPATS program emerged as the last major military aircraft competition to occur in the twentieth century. Hoping to build popular support to sway congressional delegations, contenders peppered U.S. aviation and defense magazines with major advertising appeals. Such a large production order would contribute to lower unit prices, making the plane an attractive candidate for sales to other air forces worldwide. Protocols stipulated that the winning competitor build at least 70 percent of the aircraft in the United States, but the size of the order commanded the attention of foreign manufacturers; nine American bidders eventually made JPATS proposals, all involving teams with major international partners. By the spring of 1995, Raytheon Aircraft won the JPATS contest with its Beech Mk. 2, developed with the Swiss firm of Pilatus, which had sold similar planes to a variety of foreign air arms. Projected sales of an additional 800 models of the Beech trainer to other countries underscored the significance of the export market. For some U.S. firms, teaming arrangements worked in other ways; IBM teamed up with Westland helicopters in Great Britain to develop advanced antisubmarine equipment for the EH-101 helicopter, a British and Italian design scheduled for service with British and Canadian defense forces.[5]

Among the major aerospace contractors, Boeing Corporation remained both positive and aggressive about the future. Boeing expected downturns in its various defense operations; meanwhile, sales of its commercial transports continued to generate a substantial cash flow. In addition, the company formally announced a trend-setting new airliner in 1991, the Boeing 777, scheduled to fly in mid-1994. The largest twinjet in Boeing's inventory, the 375-passenger 777 was developed as a replacement for aging wide-body transports and as competition for the newer McDonnell MD-11 and the Airbus A340. Later versions of the 777 were proposed to feature the range and passenger capacity of early models of Boeing's own 747 aircraft. To achieve its projected performance, the new 777 relied on advanced engines, a new super-efficient wing, and a variety of weight-saving alloys and composites to reduce pounds, increase payloads, and stretch its range. Boeing enlisted a number of risk-sharing partners, the principal ally being a Japanese consortium that planned to build 20 percent of the airframe.

At the same time, Boeing implemented a major change in its corporate culture. The 777 progressed as the company's first plane to be completely developed with the "design/build concept," in which prototype tooling and mock-up aircraft were essentially eliminated. Instead, the complete plane was designed and manufactured to final specifications the first time, thereby eliminating costly changes and modifications after aircraft entered service and expensive machine tools were already in place. As a means to achieve this goal, Boeing designed the 777 as a "paperless" airplane. Designers, manufacturing engineers, suppliers, and launch operators were all tied together in a comprehensive computerized network that virtually eliminated expensive and time-consuming engineering drawings and manuals. Moreover, the computer network was designed to generate detailed three-dimensional images of major components and subassemblies so that all participants could see how tooling should be built to eliminate incompatible interface situations. Such 3–D images were also able to address ergonomics, displaying maintenance operations to ensure humans could effectively accomplish them. By the time the third design cycle for the 777 ended late in the spring of 1991, the computer system had pinpointed 2,500 interface problems. In a traditional design process, such problems might not have been caught for several years. The "design/build concept" represented collective savings to the manufacturer in the tens of millions of dollars.[6]

For Boeing as well as McDonnell Douglas, the keenest competition came from overseas. During the late 1980s, at Toulouse, France, Airbus Industrie completed tooling for assembly of the A340, a four-engine, long-range jet to compete with mid-size jumbo airliners like the MD-11 and Boeing's new 777. As the A340 attained operational status (and garnered international sales) in the early 1990s, the growing controversy between Airbus Industrie and U.S. manufacturers began to take on crisis proportions. At issue were the apparent subsidies that Airbus Industrie, as an international European consortium, enjoyed from member nations. Following its formation in 1970, the Airbus consortium developed six commercial airliners. By the early 1990s, the consortium was fulfilling some 30 percent of all new airliner orders, in competition with McDonnell Douglas and Boeing. Although Boeing still led the world in sales of airliners, McDonnell Douglas dropped to third place. The United States claimed that Airbus had the benefit of European government subsidies of 60 percent to 90 percent (more than $13 billion) of program costs. European industry countered that charge by pointing to benefits that U.S. manufacturers derived from NASA and military research; American manufacturers contended that such assistance was negligible in comparison to the massive funds contributed by Airbus member nations. The United States and its European competitors held many months of talks over the subsidy issue but failed to reach a definitive compromise. American companies were intent on securing a more competitive environment over the next two decades or so, as manufacturers aimed at an estimated $600 billion market in new jet airliners. As the European community moved closer to economic integration, many Americans saw the airplane dispute as a harbinger of global competition with Europe for market share in other key technologies.[7]

In view of declining defense sales, many analysts felt that the strongest future in the aerospace market was in electronic and communications systems. During the Cold War thaw, more and more deals were struck between the United States and the new Russia. In addition to airframes and engines, a range of other aerospace components became sources of joint ventures utilizing innovative funding arrangements. One deal involved Westinghouse Electronic Systems and two European companies (Thomson CSF of France and Alenia of Italy), which agreed to collaborate in a multibillion-dollar contract to restructure Russia's outdated air traffic control system. With some estimates of total worldwide sales running as high as $10 billion, electronics for aerospace markets in the future promised to be profitable. At the same time, planning ahead to succeed in aircraft and electronics projects seemed to argue for dramatic changes in the ranks of aerospace corporations.[8]

Divestiture, Merger, and Global Partners

For American contractors, the 1990s meant a grim decision to slim down for a future of fewer contracts and scarce dollars. In some cases, the strategy included selling off some divisions to raise cash, reduce debt and interest load, and enter the new era as a leaner and meaner competitor.

A few CEOs did this despite protests from some stockholders and defense analysts. William A. Anders, a former astronaut and CEO of General Dynamics, was either a shrewd financial genius or an opportunistic plunderer, depending on the source of the opinion. A huge, successful defense contractor to the Pentagon, GD sold advanced Trident-class submarines, the Abrams battle tank, large Atlas rockets, and the internationally successful F-16 fighter plane. The company also faced a plummeting market for its military products, carried substantial debt, and had begun to lose tens of millions of dollars. Anders and his corporate advisors pulled off a series of sales that restored the company's financial status and quadrupled the stock value for long-time shareholders.

A preliminary move came early in 1992 when GD sold Cessna, considered to be a healthy unit, to Textron, arguing that GD needed to pay close attention to its "core" military business. A few months later, the momentum of divestiture picked up when GD sold off its tactical missile division to Hughes Aircraft, a company that wanted to expand its own business in this area. Given GD's historic association with the military aircraft industry, the sudden announcement that Lockheed was buying the combat airplane division—meaning the F-16—came as a shock. But the deals made by Anders represented a huge windfall for the company. "You've got to give him credit," said an aerospace analyst for an eastern bank. "He took a company with too much debt and overhead and problem contracts and either by fixing them or selling them off, he's made a lot of money for his shareholders." Anders defended his policies of "monetizing" GD by arguing that "swordmakers don't make very good or affordable ploughshares." Accordingly, he decided to sell off key segments rather than attempt to become a civil production company or continue to compete in a rapidly shrinking defense market.[9]

Among the most notable consolidations to occur was that of Martin Marietta and GE Aerospace. Martin Marietta's purchase of GE, for $3 billion, melded two of the industry's strongest teams in research and development. Together, the companies suggested formidable capabilities. GE built satellites and held a variety of Navy contracts ranging from radar and sonar equipment to computers for combat management. Martin Marietta did business in many different areas that were complementary in an R&D sense. There were its own classified spy satellites, along with Titan rockets and tactical missiles, promising an integrated launch/payload capability of considerable diversity. Other synergies seemed apparent in its computer systems, infrared sensors, and various NASA contracts.[10]

Another example of the kaleidoscopic changes of the 1990s involved the company that had begun as Vought Aeroplane back in the 1920s. After many years as a division of United Aircraft in Stamford,

Connecticut, Vought spent many more years as a component of the Ling-Temco-Vought group in Dallas, Texas. During the summer of 1992, the aircraft division of LTV was sold to a conglomerate composed of the Carlyle Group and Northrop Aviation. As the Vought Aircraft Company, this corporate survivor remained active as a subcontractor, producing vertical fins for the Boeing 757 and 767, as well as fuselage components for the 747-400. The company's past work on the Rockwell B-1 and the Northrop B-2 Stealth bomber left it with a variety of advanced machine tools and fabrication systems that management hoped to utilize for the manufacture of a new generation of air transports for the twenty-first century.[11]

During the same era of the early 1990s, Grumman's airframe design and production activity trailed off with the final deliveries and upgrades for its F-14 Tomcat and F-2C Hawkeye. From 33,700 employees in 1987, the company slimmed down to 21,000 over the next six years. The corporation's expanding business in the areas of electronics and complex systems integration for such ventures as the Joint STARS program resulted in restructuring along these lines as an electronics system specialist. In 1992, 90 percent of the company's $3.5 billion in revenues came from government programs. Analysts began to cite Grumman as a likely candidate for acquisition. During 1994, Martin Marietta and Grumman announced a merger, only to find the deal hotly contested by Northrop. Eventually, Northrop prevailed, and Grumman, long considered one of the major aircraft manufacturers, was now simply viewed as the electronics division of a larger entity. The new company, Northrop Grumman, then purchased the rest of Vought, making this combined corporate entity a major player in the defense business.[12] But bigger deals were already in the works.

Before the dust had settled from the Martin Marietta/GE agreement and the Northrop/Grumman merger, news of an even larger deal hit the headlines. Lockheed and Martin Marietta announced a corporate marriage costing an estimated $10 billion. The new aerospace giant listed combined revenues of some $23.5 billion, with products ranging from transports and the most advanced combat planes to a stable of missiles and rocket launch vehicles, as well as a myriad of electronic systems and services. Analysts had been predicting a continued run of such deals as the defense budget continued to drop, and between 1985 and 1994 the budget fell by a full one third (after inflation). Executives in the industry clearly saw merger as the way to corporate survival, and many officials in military and government agreed. By combining firms, proponents saw a means to safeguard key technologies from atrophy if one company failed to land a big contract, even though it possessed critical skills in certain areas. They argued that larger corporations could also make better utilization of production facilities and could conceivably lower unit costs. Some observers wondered if larger corporate bureaucracies might stifle internal creativity and reduce innovation in an industry dominated by just three or four huge companies.[13] Meanwhile, layoffs of personnel continued through the mid-1990s.

The dramatically altered economic status of the aerospace industry appeared with stark clarity in the AIA's figures accumulated to the end of 1995. Total sales of $107 billion were down from about $140 billion in 1991 and $123 billion in 1993, with a decline in Department of Defense spending as the major factor. From a high of $54 billion for military aircraft and missiles in 1987, defense orders dropped to $38 billion in 1995. Aerospace exports still accounted for about 7 percent to 8 percent of total American merchandise sold overseas, but the sales decline clearly hurt. All of this meant continuing layoffs, reducing the workforce to a historic low of below one million to 788,000, although the rate of decrease began to slow. The total number employed by U.S. companies who made durable goods declined from about 10 percent to about 7 percent.

For aerospace companies who could expect to hold on, there seemed reason for cautious optimism. Moody's Investors Service predicted 1996 as the start of a new upcycle in the commercial aircraft industry, with more than 15,000 new jet transports required through the year 2014. Projected geographic sales underscored the importance of the export market for American manufacturers. Based on dollar volume, Asia promised to have the strongest demand, at 34 percent of sales, with Europe accounting for another 23 percent. The North American market was expected to generate 33 percent, with the remainder coming from other corners of the globe. Always significant, the export business appeared to carry even greater weight in the future.[14]

By the mid-1990s, the financial outlook for McDonnell Douglas seemed much less bleak. Revenues in 1993 came to $14.5 billion. After struggling through a troubled R&D phase, its C-17 transport began to reach operational squadrons, and even the reduced military purchase mandated by Congress looked like it would allow some profit. The U.S. Navy's T-45 trainer, developed in cooperation with British Aerospace based on the latter's widely used Hawk series, entered service. The company received new orders for Apache helicopters, and overseas orders for fighter aircraft kept assembly lines running. The U.S. Air Force signed a $1 billion contract for Delta rocket launch vehicles, while various research and development awards totaled nearly $100 million.

McDonnell Douglas's airliner business appeared to be slightly more problematic. Some critics contended that Douglas Aircraft, which generated about one-third of the corporate revenues, remained in jeopardy, with declining orders stretching to the year 2000. Other observers countered by noting that Douglas Aircraft had made severe reductions in its cost structure during the mid-1990s, emerging, in all probability, as the industry's lowest cost producer; the break-even point for sales of its transports dropped by more than half. Douglas also continued its successful technique of taking its own and its competitors' planes as trade-ins, then placing them with smaller, emerging airlines in other countries. The corporation's financial strength also rebounded enough so that it could help finance bridge loans for customers, giving it another marketplace advantage. With the launch of a new, highly economical 100-seat airliner in 1995—the MD-95—the McDonnell Douglas corporate family hoped to remain a fixture in the aerospace industry.[15]

Boeing looked to the future with a 1993 backlog of $73.5 billion in orders, 96 percent from the civil sector. Even though as much as 20 percent of current annual revenues came from its defense sales, Boeing's future seemed strongly tied to the commercial airline business. The company recognized that technology for its own sake no longer counted as much in the customers' view, and so corporate strategy for the coming decades stressed lower costs, quicker delivery, and timely support services. Boeing also accepted the fact that it would be involved in many more projects in which it was no longer the sole or even prime contractor. The new corporate culture stressed the need for partnerships and teaming, including major contributions from foreign associates. In any case, with about 88 percent of Boeing's airliner orderbook in 1995 having overseas addresses, an international outlook was essential. For Boeing and others, the American aerospace industry functioned more and more as a global enterprise.[16]

The Asian aerospace industry continued to make gains in production capacity and in the global marketplace. Japan was sure to be an increasingly significant competitor in the civil market for components and subsystems, with strong growth in Taiwan, South Korea, and other Asian nations. Both China and Japan were positioned for rapid development in airframe production. As a result of Douglas Aircraft's collaboration in China, many MD-80 series airliners were completely assembled there; most went to Chinese operators, but in the spring of 1993, TWA bought five MD-83 planes that had been completely assembled in Shanghai. Boeing's cooperation with Chinese industry started as early as 1980 and included large assemblies and parts for the 737 and 757. In the summer of 1994, the partnership began a significant new era when Xian Aircraft Company assumed total responsibility for building the entire Boeing 737 tail assembly, requiring extensive tooling and fabrication development.

Japan-U.S. collaborations with engine manufacturers and other U.S. firms, especially Boeing and McDonnell Douglas, continued to expand in both civil and military programs. Analysts in America remained fascinated by Lockheed's arrangement to build a new Japanese fighter called the FS-X. Dating from a controversial agreement with General Dynamics, the project continued under Lockheed/Martin Marietta's direction once it had acquired GD. The plane was based on the F-16, but critics argued that the reengineered design for Japan traded away too much American combat plane technology. Other observers felt that Japan already possessed this technology, and that the joint program with Lockheed represented Japanese concessions to American political influence. In any case, the Japanese also began active conversations with foreign aerospace firms in an effort to undertake the larger role of lead contractor in the smaller passenger transports envisioned for the Asian market.[17]

Symbols, Continuity, and Conclusions

The term "Yankee technology" has often served as an automatic explanation for striking success of tools, hardware, and machines made in America. Compared to Continental examples, American technology has often exhibited original and distinctive characteristics, but the Yankee lineage has always reflected a strong European legacy. Stalwart symbols of the American frontier such as the log cabin, axe, and Pennsylvania rifle had immediate transatlantic forbears. Other familiar icons of American know-how—including Clipper ships, Conestoga wagons, and the railroad—shared this overseas heritage, and reflected the American penchant for speed, lightness, flexibility, and versatility. These characteristics reflect the insights of Richard K. Smith in his interpretation of the MTOW factor relative to American aircraft designs of the 1930s. Subsequent postwar designs featured a flexibility and versatility (stretched fuselages, commonality in design and construction among airliner and aircraft corporate families) that continued to reflect earlier generations of technological experience. The European inheritance, generously acknowledged in other areas of the history of technology, has not been similarly acknowledged in the context of American aerospace history. Having absorbed European contributions, however, American aerospace entrepreneurs often adapted and modified them into far more successful applications.

In their landmark study, *The Technical Development of Modern Aviation*, Ronald Miller and David Sawers stress the factor of keen competition in explaining the leadership of American aircraft manufacturers in the world market, along with the advantages of a large indigenous market that provided economies of scale in keeping production costs under control. To a large degree, these factors benefitted both the civil and military sectors. But there has also been another key element that helps explain the success of the American aerospace industry: the systems management procedure. Practically applied during the 1950s in the development of complex weapons systems, the idea spread throughout the industry, and enabled the United States to carry off such massive undertakings as the Apollo lunar landing program. The astute British aerospace expert Bill Gunston remarked that even advanced aviation economies like the United Kingdom's had "amazingly little" understanding of the concept, even though articles and reports about the subject had been circulating since the mid-1950s. Explaining the success of the systems approach in the United States intrigued him. "The Americans are very good at identifying abstract things and naming them, which makes subsequent progress much simpler than in countries where the problems and solutions stay ill-defined and anonymous," he wrote. "Such a concept has [had] tremendous and far-reaching implications in practice."[18]

In the process, aviation and the aerospace industry were increasingly seen to be integral parts of American life. Like most enterprises, the aerospace arena generated its own folklore and mythology. One example of the industry's rising public profile was manifest in the popular radio show "Life of Reilly" (1944–49). Starring William Bendix, the story revolved around the foibles of a prototypical industrial worker in the United States. It was notable that the program's writers felt comfortable in using the life of an airplane assembly worker. The choice not only underscored the public's interest in aviation, but also acknowledged that the industry had matured to the point where American audiences could collectively identify with it.

The aviation industry was also the source of a dictum that has worked its way into commonplace usage and will likely remain there forever: "Murphy's Law." This principle, loosely interpreted, holds that if something in a system can fail or malfunction, it will; it will also do so at the most inappropriate time. There really was a person named Murphy; as an Air Force captain in 1949, Edward A. Murphy, Jr., reported in at Edwards AFB in California under orders to investigate a mysteriously failed experiment. The procedure involved body sensors that were to monitor the brutal forces affecting Air Force test pilots. Murphy concluded that a technician had incorrectly installed all 16 of the sensors to harness fixtures, rendering them completely ineffective. "If there are two or more ways to do something," Murphy grumped, "and one of those ways can result in a catastrophe, then someone will do it." A version of this statement, reported in a subsequent press conference, caught the attention of reporters and eventually found its way into common public usage. Murphy himself eventually retired from the Air Force and continued as a reliability engineer for Hughes Helicopters. Variations on Murphy's theme continue to circulate.

There is a related theme involving the folklore surrounding "Gremlins," aggravating little creatures directly responsible for malfunctions in all sorts of complex mechanical devices. A British creation dating from World War II, references to Gremlins have appeared in the lingo of NASA astronauts, having survived into the era of space travel as multinational and mythological pests. Also in World War II, American cartoonist Robert Osborne created "Dilbert," the quintessential goofball, capable of bringing any mechanical device to an untimely, unnecessary and (usually) destructive halt. This hapless figure continued to populate instructional cartoons and machine shop conversations in the 1990s.[19]

Murphy's pessimism notwithstanding, a further theme of the industry's history has been the gritty persistence of entrepreneurs and small firms aspiring to success in the shadows of aerospace giants. Consider the complex helicopter, a business dominated by Bell, Boeing, McDonnell, and Sikorsky, with a determined Kaman still producing its own specialized designs. Enter Robinson Helicopters, organized in 1971 in Torrance, California. Frank Robinson was convinced that his design for a lightweight, two-seat helicopter could succeed. He wanted to keep its price comparable to small sport aircraft, a goal achieved by a simplified rotor control system, stubbornly basic construction, and a simple piston engine in place of much more expensive turboshaft power plants. The first production version, R22, flew in 1975, and the company had chalked up sales of 2,000 units by 1992. The R22 became a successful trainer and utility aircraft, especially with a variety of smaller countries overseas that needed helicopters but lacked the cash for larger, costlier products.

Robinson's success overseas continued to build momentum. Early in January 1993, the company announced its largest single contract to date, totaling $7 million. The deal involved the police department of Buenos Aires, Argentina, which bought 40 model R22 helicopters, along with spares, customized tools, training manuals, and instruction for both maintenance and pilot personnel.

There are other examples. Discovering a niche in an otherwise crowded marketplace likewise served Maule Aircraft in Moultrie, Georgia, whose simple, rugged, four-place planes continued to find a loyal following among the ranks of bush pilots in Alaska, Canada, and elsewhere. And while most jet engines came from huge corporations, Williams International built up a lively business during the 1950s with a compact jet engine developed especially for military cruise missiles.

Founded by a former power plant engineer at Chrysler, the company expanded into jet engines for several corporate jets, and eventually struck a deal with Rolls-Royce to help market and support a new family of such engines for worldwide markets. These and similar entrepreneurial enterprises not only proved that smaller business concerns could still compete but that they also contributed to the lively character of the aerospace marketplace. And yet, foreign influence remained significant, as Robinson Helicopters curried markets abroad, and Williams joined with Rolls-Royce on the other side of the Atlantic.[20]

The global interactions involving manufacturers of aviation and space products continued to be woven of many strands of technology, marketing, and politics. The fact that so many businesses on every continent could fabricate components, ship them to the United States, and expect them all to conform to consistent criteria for quality and fit owed much to work accomplished by the AIA. In 1975, the organization became the Secretariat of ISO/TC20, a committee having responsibility for worldwide aerospace standards. This group formulated procedures to ensure that U.S. national standards harmonized with international aerospace standards. The interdependence of aeronautic and astronautic ventures of succeeding decades vindicated the AIA's foresight in promoting such an international agenda. At the same time, the AIA realized the overseas challenge to the indigenous American industry. To enhance the nation's competitive edge, the AIA also promoted an agenda to identify "Key Technologies for the 1990s," launched in 1988. This undertaking helped identify a broad range of areas for continued research and development along with more careful integration of resources and assembly line personnel to attain "lean production" for enhanced productivity.[21]

During the dramatic restructuring of the aerospace industry in the early and mid-1990s, both of these agendas—global partnership, and the need to retain a competitive edge—remained valid. While the need for American companies to achieve competitive ability in the national and international marketplace continued, it seemed just as clear that many businesses would move into the future as foreign partnerships or find essential profit margins from sales overseas. American aviation and aerospace projects continued to emerge from workshops, and more often than not succeeded as part of a global enterprise.

APPENDIX

TABLE 1. U.S. Aircraft Production Calendar Years 1909–74 (Number of Aircraft)

Year Ending December 31	Total[a]	Military	Civil
1909	N.A.	1	N.A.
1910	N.A.	—	N.A.
1911	N.A.	11	N.A.
1912	45	16	29
1913	43	14	29
1914	49	15	34
1915	178	26	152
1916	411	142	269
1917	2,148	2,013	135
1918	14,020	13,991	29
1919	780	682	98
1920	328	256	72
1921	437	389	48
1922	263	226	37
1923	743	687	56
1924	377	317	60
1925	789	447	342
1926	1,186	532	654
1927	1,995	621	1,374
1928	4,346	1,219	3,127

(Continued)

TABLE 1. U.S. Aircraft Production (cont.)
Calendar Years 1909–74 (Number of Aircraft)

Year Ending December 31	Total[a]	Military	Civil
1929	6,193	677	5,516
1930	3,437	747	2,690
1931	2,800	812	1,988
1932	1,396	593	803
1933	1,324	466	858
1934	1,615	437	1,178
1935	1,710	459	1,251
1936	3,010	1,141	1,869
1937	3,773	949	2,824
1938	3,623	1,800	1,823
1939	5,856	2,195	3,661
1940	12,813	6,028	6,785
1941	26,289	19,445	6,844
1942	47,675	47,675	—
1943	85,433	85,433	—
1944	95,272	95,272	—
1945	48,912	46,865	2,047
1946	36,418	1,417	35,001
1947	17,739	2,122	15,617
1948	9,838	2,536	7,302
1949	6,137	2,592	3,545
1950	6,200	2,680	3,520
1951	7,532	5,055	2,477
1952	10,640	7,131	3,509
1953	13,112	8,978	4,134
1954	11,478	8,089	3,389
1955	11,484	6,664	4,820
1956	12,408	5,203	7,205
1957	11,943	5,198	6,745
1958	10,938	4,078	6,860

TABLE 1. U.S. Aircraft Production (cont.)
Calendar Years 1909–74 (Number of Aircraft)

Year Ending December 31	Total[a]	Military	Civil
1959	11,076	2,834	8,242
1960	10,237	2,056	8,181
1961	8,936	1,582	7,354
1962	9,213	1,975	7,238
1963	10,143	1,970	8,173
1964	12,517	2,439	10,078
1965	15,489	2,806	12,683
1966	20,283	3,609	16,674
1967	18,993	4,481	14,512
1968	19,362	4,440	14,922
1969	17,149	3,644	13,505
1970	11,161	3,085	8,076
1971	10,390	2,232	8,158
1972	12,569	1,993	10,576
1973	15,952	1,243	14,709
1974	16,325[E]	1,000[E]	15,325

SOURCE: Aerospace Industries Association, company reports; General Aviation Manufacturers Association, company reports; Department of Defense.

[a]Excludes aircraft produced for the Military Assistance Program.

[E]Estimate.

TABLE 2. U.S. Aircraft Production–Civil
Calendar Years 1969–94

Year	TOTAL	Domestic Shipments			Export Shipments		
		Transports[a]	Helicopters	General Aviation	Transports	Helicopters	General Aviation
1969	13,505	332	282	9,996	182	252	2,461
1970	8,076	127	150	5,246	184	332	2,037
1971	8,158	50	171	5,900	173	298	1,566
1972	10,576	79	319	7,702	148	256	2,072
1973	14,709	143	342	10,482	151	428	3,163
1974	15,326	91	433	9,903	241	395	4,263
1975	15,251	127	528	10,804	188	336	3,268
1976	16,429	64	442	12,232	158	315	3,218
1977	17,913	54	527	13,441	101	321	3,469
1978	18,962	130	536	14,346	111	368	3,471
1979	18,460	176	570	13,177	200	459	3,878
1980	13,634	150	841	8,703	237	525	3,178
1981	10,916	132	619	6,840	255	453	2,617
1982	5,085	111	333	3,326	121	254	940
1983	3,356	133	187	2,172	129	216	519
1984	2,999	102	143	2,013	83	233	425
1985	2,691	126	247	1,545	152	137	484
1986	2,156	171	120	1,031	159	210	464
1987	1,800	187	116	598	170	242	487
1988	1,949	206	103	500	217	280	643
1989	2,448	138	221	225	260	294	1,310
1990	2,268	215	254	335	306	349	809
1991	2,181	204	253	487	385	318	534
1992	1,790	180	112	541	387	212	358
1993	1,630	130	83	631	278	173	333
1994	1,545	87	154	543	224	154	385

SOURCE: Aerospace Industries Association, based on company reports; General Aviation Manufacturers Association; Department of Commerce, International Trade Administration.
NOTE: Tables 2 through 6 taken from Aerospace Industries Association, *Aerospace Facts and Figures, 1995–96* (Washington, D.C.: AIA, 1995).
[a] Prior to 1976, includes the C-130 military transport.

TABLE 3. U.S. Aircraft Production–Military
Calendar Years 1969–94

Year	TOTAL	U.S. Military Agencies	Exports		
			Total	FMS[a]	Direct[b]
1969	4,290	3,644	646	NA	NA
1970	3,720	3,085	635	NA	NA
1971	2,914	2,232	682	NA	NA
1972	2,530	1,993	537	124	413
1973	1,821	1,243	578	129	449
1974	1,513	799	714	365	349
1975	1,779	844	935	525	410
1976	1,318	625	693	518	175
1977	1,134	454	680	408	272
1978	996	467	529	256	273
1979	837	531	306	203	103
1980	1,047	625	422	194	228
1981	1,062	703	359	215	144
1982	1,159	690	469	68	401
1983	1,053	766	287	70	217
1984	936	561	375	71	304
1985	919	643	276	134	142
1986	1,107	708	399	110	289
1987	1,210	725	485	133	352
1988	1,305	687	618	138	480
1989	1,261	614	647	92	555
1990	1,053	664	387	99	289
1991	911	556	355	94	261
1992	753	422	331	122	209
1993	955	437	518	146	372
1994	755	409	346	63	283

SOURCE: Aerospace Industries Association, based on USAF, USN, and USA survey responses; Department of Commerce, International Trade Administration.
[a]Foreign Military Sales, through Department of Defense. Also includes acceptances of NATO AWACS aircraft.
[b]Military aircraft exported via commercial contracts, directly from manufacturers to foreign governments.
NA Not available.

TABLE 4. Total U.S. Exports and Exports of Aerospace Products
Calendar Years 1964–92
(Millions of dollars)

Year	TOTAL Exports of U.S. Merchandise[a]	Exports of Aerospace Products				
		TOTAL	Percent of total U.S. Exports	Civil		Military
				Total	Transports	
1964	$25,690	$1,608	6.3%	$764	$211	$844
1965	26,699	1,618	6.1	854	353	764
1966	29,379	1,673	5.7	1,035	421	638
1967	30,934	2,248	7.3	1,380	611	868
1968	34,063	2,994	8.8	2,289	1,200	705
1969	37,332	3,138	8.4	2,027	947	1,111
1970	43,176	3,405	7.9	2,516	1,283	889
1971	44,087	4,203	9.5	3,080	1,567	1,123
1972	49,854	3,795	7.6	2,954	1,119	841
1973	71,865	5,142	7.2	3,788	1,664	1,354
1974	99,437	7,095	7.1	5,273	2,655	1,822
1975	108,856	7,792	7.2	5,324	2,397	2,468
1976	116,794	7,843	6.7	5,677	2,468	2,166
1977	123,182	7,581	6.2	5,049	1,936	2,532
1978	145,847	10,001	6.9	6,018	2,558	3,983
1979	186,363	11,747	6.3	9,772	4,998	1,975
1980	225,566	15,506	6.9	13,248	6,727	2,258
1981	238,715	17,634	7.4	13,312	7,180	4,322
1982	216,442	15,603	7.2	9,608	3,834	5,995
1983	205,639	16,065	7.8	10,595	4,683	5,470
1984	223,976	15,008	6.7	9,659	3,195	5,350
1985	218,815	18,725	8.6	12,942	5,518	5,783
1986	227,159	19,728	8.7	14,851	6,276	4,875
1987	254,122	22,480	8.8	15,768	6,377	6,714
1988	322,426	26,947	8.4	20,298	8,766	6,651
1989	363,812	32,111	8.8	25,619	12,313	6,492
1990	393,592[r]	39,083	9.9	31,517	16,691	7,566
1991	421,730[r]	43,788	10.4	35,548	20,881	8,239
1992	448,115	45,018	10.0	36,904	22,379	8,114

SOURCE: Bureau of the Census, Foreign Trade Division; Aerospace Industries Association, based on data from International Trade Administration.
NOTE: International trade reported using Harmonized Tariff Schedules after 1988.
[a]Includes DOD shipments and undocumented exports to Canada, free alongside ship basis.
[r]Revised.

TABLE 5. Aerospace Industry Sales by Customer
Calendar Years 1978–94
(Millions of dollars)

		Aerospace Products and Services				Related Products and Services
			U.S. Government			
Year	TOTAL SALES	Total	Dept. of Defense	NASA and Other Agencies	Other Customers	
Current Dollars						
1978	$37,702	$30,889	$15,533	$3,151	$12,205	$6,813
1979	45,420	37,705	18,918	3,453	15,334	7,715
1980	54,697	45,878	22,795	4,106	18,977	8,819
1981	63,974	53,090	27,244	4,709	21,137	10,884
1982	67,756	56,366	34,016	4,899	17,451	11,390
1983	79,975	66,646	41,558	5,910	19,178	13,329
1984	83,486	69,572	45,969	6,063	17,540	13,914
1985	96,571	80,476	53,178	6,262	21,036	16,095
1986	106,183	88,486	59,161	6,236	23,089	17,697
1987	110,008	91,673	61,817	6,813	23,043	18,335
1988	114,562	95,468	61,327	7,899	26,242	19,094
1989	120,534	100,445	61,199	9,601	29,645	20,089
1990	134,375	111,979	60,502	11,097	40,379	22,396
1991	139,248	116,040	56,619	11,739	48,379	23,208
1992	138,591	115,493	52,202	12,408	50,882	23,099
1993	123,416	102,847	46,441	12,267	44,139	20,569
1994	111,223	92,686	44,013	11,936	36,737	18,577

SOURCE: Aerospace Industries Association.

TABLE 6. U.S. Total and Aerospace Foreign Trade
Calendar Years 1964–94
(Millions of dollars)

Year	Total U.S. Merchandise Trade			Aerospace		
	Trade Balance	Exports	Imports	Trade Balance	Exports	Imports
1964	$7,006	$25,690	$18,684	$1,518	$1,608	$90
1965	5,334	26,699	21,366	1,459	1,618	139
1966	3,837	20,379	25,542	1,370	1,673	303
1967	4,122	30,934	26,812	1,961	2,248	287
1968	837	34,063	33,226	2,661	2,994	333
1969	1,289	37,332	36,043	2,831	3,138	307
1970	3,225	43,176	39,952	3,097	3,405	308
1971	(1,476)	44,087	45,563	3,830	4,203	373
1972	(5,729)	49,854	55,583	3,230	3,795	565
1973	2,390	71,865	69,476	4,360	5,142	782
1974	(3,084)	99,437	103,321	6,350	7,095	745
1975	9,551	108,856	99,303	7,045	7,792	747
1976	(7,820)	116,794	124,614	7,267	7,843	576
1977	(28,353)	123,182	151,534	6,850	7,581	731
1978	(30,205)	145,847	176,052	9,058	10,001	943
1979	(23,922)	186,363	210,285	10,123	11,747	1,624
1980	(19,696)	225,566	245,262	11,932	15,506	3,554
1981	(22,267)	238,715	260,982	13,134	17,634	4,500
1982	(27,510)	216,442	243,952	11,035	15,603	4,568
1983	(52,409)	205,639	258,048	12,619	16,065	3,446
1984	(106,703)	223,976	330,678	10,082	15,008	4,926
1985	(117,712)	218,815	336,526	12,593	18,725	6,132
1986	(138,279)	227,159	365,438	11,826	19,720	7,902
1987	(152,119)	254,122	406,241	14,575	22,480	7,902
1988	(118,526)	322,426	440,952	17,860	26,947	9,087

TABLE 6. U.S. Total and Aerospace Foreign Trade (cont.)
Calendar Years 1964–94
(Millions of dollars)

Year	Total U.S. Merchandise Trade			Aerospace		
	Trade Balance	Exports	Imports	Trade Balance	Exports	Imports
1989	(109,399)	363,812	473,211	22,083	32,111	10,028
1990	(101,718)	393,592	495,311	27,282	39,083	11,801
1991	(66,723)	421,730	480,453	30,785	43,788	13,003
1992	(84,501)	448,164	532,665	31,356	45,018	13,662
1993	(115,568)	465,091	580,659	27,235	39,418	12,183
1994	(151,308)	512,521	663,829	25,010	37,373	12,363

SOURCE: Bureau of the Census, Foreign Trade Division, and Aerospace Industries Association, based on data from International Trade Administration.

CHRONOLOGY

1903	Orville Wright completes world's first powered, sustained, and controlled flight in a heavier-than-air airplane.
1907–1909	Glenn Curtiss, the Wright brothers, and others organize manufacturing companies; Wrights sign European orders and conclude foreign agreements.
1908	U.S. Army signs contract for Wright biplane.
1913	First college-level courses in aero engineering subjects offered at University of Michigan and at MIT.
1914	World War I in Europe; foreign orders for planes subsequently placed in the U.S.
1915	National Advisory Committee for Aeronautics (NACA) is founded.
1917	U.S. declares war on Germany; U.S. Congress appropriates $640 million to build more than 20,000 aircraft.
	Aircraft Manufacturers Association (AMA) formed, as first builders' group to gather statistics and provide a coordinating body.
1918	U.S. Air Mail inaugurated; provides example of civil and commercial use of aviation.
	End of World War I.
1919	Aeronautical Chamber of Commerce founded as successor to AMA.
	U.S. Navy NC-4 flying boat makes first flight across the Atlantic Ocean.
1925	Air Mail Act (the Kelly Bill) shifts airmail from government planes to commercial contractors; forms basis for the air transport industry.

1926	Robert Goddard successfully launches the first liquid propellant rocket.
	Air Commerce Act establishes safety regulations; investors and insurance underwriters more inclined to invest in aviation.
1927	Charles Lindbergh completes first nonstop solo flight across Atlantic Ocean.
1928	During the 1920s, NACA research represents significant contributions to aeronautics, including the NACA cowling design in 1928.
1926–1930	The Daniel Guggenheim Foundation for the Promotion of Aeronautics makes fundamental contributions in meteorological studies, blind flying, aero engineering education, and other aviation-related areas.
1920s	During the decade, general aviation planes are produced; builders such as Beech, Cessna, and Piper appear.
1933	Boeing 247 makes first flight.
1934	U.S. War Department forms GHQ Air Force, providing significant autonomy.
	Douglas DC-2 demonstrates advanced performance in the MacRobertson Air Race, England to Australia.
1937	Frank Whittle (Britain) completes successful demonstration of a practical jet engine.
1938	Orders for U.S. military planes by Britain and other European nations stimulate expansion of America's aircraft industry.
	National Aerospace Standards Committee established to develop technical specifications for items designed for aviation products.
1939	Heinkel HE 178 (Germany) makes first successful flight by a jet airplane.
	Germany invades Poland; Britain and France declare war on Germany.
	Igor Sikorsky makes successful demonstration of Vought-Sikorsky VS-300 helicopter with main and tail rotor, establishing a standard pattern for subsequent helicopter designs.
1941	Lend-Lease Act is passed by Congress, committing the U.S. to accelerated production of military aircraft.

United States Army Air Force is formed.

Pearl Harbor attacked by Japanese carrier aircraft; U.S. enters World War II.

1942 Pioneer rocket firms like Reaction Motors (1941) and Aerojet are founded in the U.S.

German V-2 ballistic missiles launched against targets in France and Britain.

Bell P-59, first U.S. jet, flies with British-designed engine.

Four-engine Douglas C-54 military transport (based on DC-4) enters service and becomes major Allied long-range transport in World War II; although unpressurized, establishes major conceptual trend for postwar transports.

1943 First flight of Lockheed Model L-049 Constellation; the large, four-engine, pressurized air transport represents U.S. leadership in such designs that dominate early postwar airline routes, both American and foreign flag lines.

1945 U.S. industry produces unprecedented quantities of aircraft, totaling more than 300,000 by war's end.

Project Paperclip; American military program to locate and recruit leading German aviation and rocket experts to work in the U.S.

Atom bomb dropped on Japan; end of World War II.

Aeronautical Chamber of Commerce of America becomes Aircraft Industries Association, with focus on supporting U.S. aviation industry needs.

1946 Bell Model 47 receives first commercial helicopter certificate issued by U.S.

Beechcraft Bonanza represents robust hopes and new designs of postwar general aviation manufacturers.

1947 National Security Act makes U.S. Air Force a separate military branch.

Captain Charles Yeager breaks speed of sound in Bell XS-1 rocket-powered plane.

1952 de Havilland Comet, world's first jet airliner, enters service with BOAC.

1954 Comet jet liners withdrawn from service after series of crashes.

1955	Cessna 172 introduced; exemplifies success of light, four-place, fixed gear general aviation planes.
1957	Cooperation with USAF Air Materiel Command and MIT to develop an automatic programmed tool system for machining complex aircraft and missile parts.
	Soviet Union launches Sputnik, world's first artificial satellite.
1958	Explorer 1, first U.S. satellite, successfully launched.
	First flight of McDonnell FH-1 Phantom II; represents success in mass production of aerodynamically complex fighters that exemplify electronic warfare, versatility, long service life, and export sales.
	National Aeronautics and Space Administration (NASA) supersedes NACA.
	Boeing 707 jet airliner enters service.
1959	Aerospace Industries Association; formerly the Aircraft Industries Association, the name change reflects new space age activities.
	Atlas intercontinental ballistic missiles become operational.
1960	U.S. launches first early warning satellite (Midas 1); first weather satellite (Tiros 1); first of all-weather navigation satellites (Transit 1-B series) to provide data for submarine-launched missiles.
	Piper Pawnee represents standard configuration for postwar ag-plane designs.
1961	President John F. Kennedy announces national goal of a manned lunar landing within the decade.
1962	John Glenn becomes first American to orbit the earth.
	Telstar 1, the first transatlantic satellite television relay, is placed in orbit.
1963	First flight of Lear 23, a design that spurs the subsequent success of corporate jets.
1965	Lufthansa orders the Boeing 737, the first foreign airline to launch a U.S. jet transport.

Douglas DC-9 inaugural flight; Boeing and Douglas twin-jets become world sales leaders as smaller jets rapidly spread jet airline travel.

1969 Boeing Model 747 makes inaugural flight; first of the new generation of wide-body transports, or "jumbo-jets."

Neil Armstrong and Ed Aldrin make first lunar landing, carried by the Apollo-Saturn vehicle.

1970 Airbus Industrie is established in Europe as a consortium to develop jet airliners in competition with the U.S.

General Aviation Manufacturers Association is formed.

1974 Airbus Industrie A-300 wide-body transport enters service.

1975 Belgium, the Netherlands, Norway, and Denmark form consortium to produce a common NATO fighter, the General Dynamics F-16.

AIA becomes secretariat of ISO/TC 20, the international committee responsible for developing worldwide standards for the manufacture of aerospace products.

1981 NASA's space shuttle *Columbia* makes first launch.

First flight of Lockheed F-117 "Stealth" combat plane.

1984 President Ronald Reagan endorses NASA plans to put a permanent space station into earth orbit; relies on international collaboration for basic components.

1989 Bell/Boeing V-22 Osprey tilt-rotor aircraft makes full conversion to the airplane mode while in full flight.

Berlin Wall is demolished, symbolizing collapse of Soviet Union's power and demise of the Cold War.

1991 Coalition forces launch Desert Storm to defeat Iraqi forces; impressive demonstrations of U.S. electronic warfare and Stealth technology.

Strategic Arms Reduction Talks (START) accelerates dismantling of nuclear missiles, including ICBMs.

1992 Lockheed buys F-16 manufacturing division from General Dynamics, typical of merger trends as U.S. aerospace corporations experience major downsizing with restructuring.

1994	Insurance/liability reform bill to benefit general aviation industry passes Congress.
	Mergers continue, as Northrop buys Grumman; Lockheed merges with Martin Marietta.
1995	NASA space shuttle docks with Russian space station in earth orbit.
1997	Boeing absorbs McDonnell Douglas in historic merger.
2000	American and Russian astronauts occupy the International Space Station for the first time.

NOTES

PREFACE

1. See, for example, David D. Lee, "Herbert Hoover and the Golden Age of Aviation," in William Leary, ed., *Aviation's Golden Age: Portraits from the 1920s and 1930s* (Iowa City: University of Iowa Press, 1989), 127–47, and Ellis Hawley, "Three Facets of Hooverian Associationalism: Lumber, Aviation, and Movies, 1921–1930," in Thomas McCraw, ed., *Regulation in Perspective* (Cambridge: Harvard University Press), 95–123. For the military dimension see Jacob A. Vander Meulen, *The Politics of Aircraft: Building an American Military Industry* (Lawrence: University Press of Kansas, 1991).

CHAPTER 1

1. Marvin McFarland, "Wilbur and Orville Wright: Seventy-five Years After," in Richard P. Hallion, ed., *The Wright Brothers: Heirs of Prometheus* (Washington: Smithsonian Institution Press, 1978), 21. For a definitive biography, see Tom Crouch, *The Bishop's Boys: A Life of Wilbur and Orville Wright* (New York: W. W. Norton, 1989). For a careful analysis of the Wrights' development of their early gliders and first powered machine, see Peter Jakab, *Visions of a Flying Machine: The Wright Brothers and the Process of Invention* (Washington, D.C.: Smithsonian Institution Press, 1990).
2. Roger E. Bilstein, "The Airplane, the Wrights, and the American Public," in Hallion, ed., *Heirs*, 39–51.
3. Ibid.; Henry H. Arnold, *Global Mission* (New York: Harper, 1942), 14; Fred C. Kelly, ed., *Miracle at Kitty Hawk: The Letters of Wilbur and Orville Wright* (New York: Farrar, Straus, and Young, 1951), 133.
4. Grover Loening, *Take Off into Greatness: How American Aviation Grew So Big So Fast* (New York: Putnam, 1968), 26–29; Crouch, *Bishop's Boys*, 291–359.

5. Crouch, *Bishop's Boys*, 387; 362–94, 402–10; Fred Howard, *Wilbur and Orville: A Biography of the Wright Brothers* (New York: Alfred A. Knopf, 1987), 224–27; Fred C. Kelly, *The Wright Brothers: A Biography Authorized by Orville Wright* (New York: Harcourt Brace, 1943), 164–66.

6. Grover Loening, *Our Wings Grow Faster* (New York: Doubleday, Doran, 1935), 44, 50; Crouch, *Bishop's Boys*, 411–12, 415–18, 425, 435, 440–42, 447–51.

7. Welman Austin Shrader, *Fifty Years of Flight: A Chronicle of the Aviation Industry in America, 1903–1953* (Cleveland: Eaton Manufacturing Company, 1953), 17–23; John B. Rae, *Climb to Greatness: The American Aircraft Industry, 1920–1960* (Cambridge: MIT Press, 1968), 32–35; John Evangelist Walsh, *One Day at Kitty Hawk: The Untold Story of the Wright Brothers and the Airplane* (New York: Crowell, 1975), 420–23; Crouch, *Bishop's Boys*, 455–66, 470–79.

8. C.R. Roseberry, *Glenn Curtiss: Pioneer of Flight* (New York: Doubleday, 1972), 1–82, offers the fullest account of Curtiss's early career. Ballooning in America enjoyed a long history, dating from the late eighteenth century, and military observation balloons were used by both Union and Confederate forces during the Civil War. An encyclopedic survey is Tom Crouch, *The Eagle Aloft: Two Centuries of the Balloon in America* (Washington, D.C.: Smithsonian Institution Press, 1983).

9. Roseberry, *Glenn Curtiss*, 83–203; Louis Casey, *Curtiss: The Hammondsport Era, 1907–1915* (New York: Crown, 1981), 42, passim. One of the Curtiss 1908 advertisements is reproduced in the latter, p. 9.

10. Bilstein, "Airplane," in Hallion, *Heirs*, 42–44; Roseberry, *Glenn Curtiss* 281–307.

11. Interview of Frank Coffyn, housed in the Oral History Collections, Columbia University, New York City. Cited hereafter as OHC. Bilstein, "Airplane," in Hallion, *Heirs*, 48–50; Donald Douglas, OHC. On the career of Glenn Martin, Douglas, and other colorful aeronautical entrepreneurs of the era, see Wayne Biddle, *Barons of the Sky: From Early Flight to Strategic Warfare—The Story of the American Aerospace Industry* (New York: Simon and Schuster, 1991).

12. Casey, *Curtiss*, 176–77; Roseberry, *Glenn Curtiss*, 365–66, 397.

13. Harold E. Morehouse, "Flying Pioneers Biographies" summary of "Thomas Brothers," from an unpublished manuscript in the National Air and Space Museum Archives (n.p., n.d.). Morehouse was an early employee at Dayton-Wright and with other firms after World War I. His "Biographies" files represent an extensive compilation for a proposed book. If not always accurate, they constitute an informative collection on early flight.

14. Shrader, *Fifty Years*, 10–16. Equipment miscellany is culled from exhibit materials at the National Air and Space Museum, especially the artifact lists and working script for the "Early Flight" gallery, NASA Archives.

15. Tom Crouch, *Bleriot XI: The Story of a Classic Aircraft* (Washington, D.C.: Smithsonian Institution Press, 198), 61–64. Walsh, *One Day*, 391–92; Loening, *Wings*, 21–30, 48–55, 61–62.

16. Alex Roland, *Model Research: The National Advisory Committee for Aeronautics, 1915–1958*, vol. I (Washington, D.C.: U.S. Government Printing Office, 1985), 1–22, and vol. II, pages 394 and 571. The latter volume is essentially a collection of selected documents, and reprints the NACA Charter as well as the British committee report.

17. Ibid., I, 22–49; James R. Hansen, *Engineer in Charge: A History of the Langley Aeronautical Laboratory, 1917–1958* (Washington, D.C.: U.S. Government Printing Office, 1987), 8–22.

18. Dominick Pisano, with Thomas Dietz, Joanne Gernstein, and Karl Schneide, *Legend, Memory, and the Great War in the Air* (Seattle: University of Washington Press, 1992), 88–91, 96, 98. See also Philip S. Dickey III, *The Liberty Engine, 1918–1942* (Washington, D.C.: Smithsonian Institution Press, 1968), passim. Dickey put the production total at 20,478 by February 1919, with peak production of 3,878 per month in October 1918 (p. 91).

19. Loening, *Wings*, 45; Crouch, *Bishop's Boys*, 246, 360–63, 402–03.

20. Roseberry, *Glenn Curtiss*, 381–93; Crouch, *Bishop's Boys*, 463, 489–96; Rae, *Climb*, 5; Loening, *Take Off*, 62–63, 89, 114–15.

21. Loening, *Wings*, 63–72.

22. G. R. Simonson, "The Demand for Aircraft and the Aircraft Industry, 1907–1958," *Journal of Economic History*, vol. 20 (September 1960), 362–63.

23. Casey, *Curtiss*, passim.

24. U.S. Bureau of the Census, *Census of Manufacturers, 1914* (Washington, D.C.: U.S. Government Printing Office, 1917), 26, 568; U.S. Civil Aeronautics Authority, *Air Commerce Bulletin*, vol. I (Washington, D.C.: U.S. Government Printing Office, 1929), 6.

25. H. B. Hickam, "The Truth about Our Aeroplane Record," *Current History* 15 (October 1921): 48; "Senate Investigation of the War Department," *Aerial Age* 7 (April 15, 1918), 262.

26. *Air Commerce Bulletin*, vol. I, 6; Victor Selden Clark, *History of Manufacturers in the United States*, vol. II (New York: McGraw-Hill, 1929), 315.

27. Irving Brinton Holley, *Ideas and Weapons* (New Haven: Yale University Press, 1953), 45–60, 141–46; Dickey, *Liberty Engine*, passim; Alfred A. Goldberg, ed., *A History of the United States Air Force, 1907–1957* (Princeton: Van Nostrand, 1957), 18; David Anderton, *The History of the*

U.S. Air Force (New York: Crescent Books, 1981), 26; Shrader, *Fifty Years*, 20; Rae, *Climb*, 222–24. In Rae's tabulations, 41,953 engines are listed separately. The value of airframes totaled $113.7 million; that of engines came to $244.8 million. Rae also tabulated 41,387 balloons and airships valued at $7.1 million. The term "aircraft industry" can be misleading, since automotive and miscellaneous interests delivered over 4,000 of the 13,894 airframes; in the case of engines, the automobile and kindred industries delivered more than 27,000 of the 41,953 total units.

28. Elsbeth E. Freudenthal, *The Aviation Business: From Kitty Hawk to Wall Street* (New York: Vanguard, 1940); Rae, *Climb*, 1–3; Jacob Vander Meulen, *The Politics of Aircraft: Building an American Military Industry* (Lawrence: University Press of Kansas, 1991), 8–40.

CHAPTER 2

1. The competitive element is implicit in Terry Gwynn-Jones, *Farther and Faster: Aviation's Adventuring Years, 1909–1939* (Washington, D.C.: Smithsonian Institution Press, 1991), especially chapter 6, "For National Honor," 103–38.

2. Gwynn-Jones, *Farther and Faster*, 114–38. See also Paul O'Neil, *Barnstormers and Speed Kings* (Alexandria, Va.: Time-Life Books, 1981), passim.

3. On the Ford experience, see William M. Leary, "Henry Ford and Aeronautics during the 1920s," in Leary, ed., *Aviation's Golden Age: Portraits from the 1920s and 1930s* (Iowa City: University of Iowa Press, 1989), 1–17. John B. Rae, *Climb to Greatness* (Ch. 2, Note 12), profiles many of the early airframe and engine manufacturers. See also Wayne Biddle, *Barons of the Sky*, passim, for pointed commentary on Douglas, Lockheed, and Northrop (Ch. 1, Note 11). The origins of Boeing and Grumman are covered in E.E. Bauer, *Boeing in Peace and War* (Enumclaw, Wash.: TABA Publishing, 1991), 4–54; Bill Gunston, *Grumman: Sixty Years of Excellence* (New York: Orion Books, 1988), 8–25.

4. For origins of airmail see William M. Leary, *Aerial Pioneers: The U.S. Air Mail Service, 1918–1927* (Washington, D.C.: Smithsonian Institution Press, 1985). For socioeconomic aspects of airmail as well as the evolution of light airplanes, see Roger E. Bilstein, *Flight Patterns: Trends of Aeronautical Development in the United States, 1918–1929* (Athens: University of Georgia Press, 1983), 29–56, 59–96; airline developments are detailed in R.E.G. Davies, *Airlines of the United States since 1914* (Washington, D.C.: Smithsonian Institution Press, 1982).

5. For a convenient summary of military trends, see Alfred Goldberg, ed., *A History of the United States Air Force, 1907–1957* (Princeton: Van Nostrand, 1957), 1–45. See also Robert T. Finney, *History of the Air*

Corps Tactical School, 1920–1940 (Washington, D.C.: Center for Air Force History, 1992).

6. National Aeronautics and Space Administration, *Progress in Aircraft Design since 1903* (Washington, D.C.: U.S. Government Printing Office, 1974), 30–31; Ray Wagner, *American Combat Planes*, 3rd ed. (Garden City, N.Y.: Doubleday & Company, 1982), passim.

7. NASA, *Progress*, 32, 36; Wagner, *Combat Planes*, passim; William Trimble, *Admiral William Moffett: Architect of Naval Aviation* (Washington, D.C.: Smithsonian Institution Press, 1994), passim.

8. For a useful summary of the civil and military aviation business, see Irving Brinton Holley, Jr., *Buying Aircraft: Material Procurement for the Army Air Forces* (Washington, D.C.: U.S. Government Printing Office, 1964), 6–22.

9. Grover Loening, *Our Wings Grow Faster* (Garden City, N.Y.: Doubleday, Doran and Company, 1935) 108–9, 116; Jacob Vander Meulen, *The Politics of Aircraft: Building an American Military Industry* (Lawrence: University Press of Kansas), 65–73, 106–9; Holley, *Buying Aircraft*, 10–11, and fn. 13. Holley notes the problem of rectifying many statistics of the era, since some compilers listed units produced, while others listed units sold including inventory items. Export sales were often counted twice.

10. Victor Selden Clark, *History of Manufacturers in the U.S.*, vol. 2 (New York: McGraw-Hill, 1929), 338; Welman Austin Shrader, *Fifty Years of Flight* (Cleveland: Eaton Manufacturing Company, 1953), 20; U.S. Bureau of the Census, *Historical Statistics of the United States, 1789–1945* (Washington, D.C.: U.S. Government Printing Office, 1949), 412; Bilstein, *Patterns*, 127–28.

11. Bilstein, *Patterns*, 128–30; Holley, *Buying Aircraft*, 6–8; *Aviation Week and Space Technology: 75th Anniversary Issue*, 135 (August 12, 1991): 176–77, 206–7. On Edwin Link's career and the simulator business, see Lloyd Kelly, *The Pilot Maker* (New York: Grosset and Dunlap, 1970).

12. Rae, *Climb*, 38–51; Shrader, *Fifty Years*, 24–44. See also, G.R. Simonson, ed., *The History of the American Aircraft Industry: An Anthology* (Cambridge: MIT Press, 1968), chapter two.

13. Loening, *Wings*, 125–40, 169–72; Grover Loening, *Amphibian: The Story of the Loening Biplane* (Greenwich, Conn.: New York Graphic Society, 1973), 180.

14. Rae, *Climb*, 79–87; Simonson, *History*, passim; Vander Meulen, *The Politics of Aircraft*, 45–47; 90–102; 114–16.

15. For a general discussion, see William Glenn Cunningham, *The Aircraft Industry: A Study in Industrial Location* (Los Angeles: Lorrin L. Morrison, 1951). On the light plane business in Wichita, see Rae, *Climb*, 15–16. Later analyses have given much more emphasis to the initiative of local entrepreneurs and the availability of risk capital. See, for example,

Ann Markusen et al., *The Rise of the Gunbelt: The Military Remapping of Industrial America* (New York: Oxford University Press, 1991). This work focuses on the aerospace industry. The influence of local partisans and availability of capital in southern California has been carefully analyzed by Joseph E. Libby, "To Build Wings for the Angels: Los Angeles and Its Aircraft Industry, 1890–1936" (Ph.D. diss., University of California—Riverside, December 1990). Dr. Libby graciously shared his insights with the author.

16. Bilstein, *Patterns*, 100–103. For a penetrating assessment of the NACA during this era, see Alex Roland, *Model Research: The National Advisory Committee for Aeronautics*, vol. 1 (Washington, D.C.: U.S. Government Printing Office, 1985). His comments on the cowling development appear pp. 113–17, and 352, fn. 44. Langley's evolution is detailed in James R. Hansen, *Engineer in Charge: A History of the Langley Aeronautical Laboratory, 1917–1958* (Washington, D.C.: U.S. Government Printing Office, 1987). Weick's own recollections appear in Fred E. Weick and James R. Hansen, *From the Ground Up: The Autobiography of an Aeronautical Engineer* (Washington, D.C.: Smithsonian Institution Press, 1988).

17. Bilstein, *Patterns*, 110–14; Robert Weeks, "The First Fifty Years: Department of Aeronautical and Astronautical Engineering, University of Michigan" (Ann Arbor, 1964), passim. The latter is an informal, typescript history, generously shared after the author's enquiry; copy in author's files. On the role of the Guggenheim Foundation, see Richard P. Hallion, *Legacy of Flight: The Guggenheim Contribution to American Aviation* (Seattle: University of Washington Press, 1977); the recruitment of von Karman is told in Paul A. Hanle, *Bringing Aerodynamics to America* (Cambridge: MIT Press, 1982).

18. For an instructive survey of these essential trends, see Ronald Miller and David Sawers, *The Technical Development of Modern Aviation* (New York: Praeger, 1970), 53–86. On Boeing planes, see Harold Mansfield, *Vision: A Saga of the Sky* (New York: Duell, Sloan, and Pearce, 1956), 44–46. See also, Rae, *Climb*, 58–63; Peter W. Brooks, *The Modern Airliner: Its Origins and Development* (London: Putnam, 1961); and Howard Wolko, *In the Cause of Flight: Technologists of Aeronautics* (Washington, D.C.: Smithsonian Institution Press, 1981). For a useful summary by a contemporaneous observer, see Edward P. Warner, *Technical Development and Its Effect on Air Transportation* (York, Pa.: Maple Press, 1937). Eric Schatzenberg, "Ideology and Technical Choice: The Decline of the Wooden Airplane in the United States, 1920–1945," *Technology and Culture* 35 (January 1994), 34–69, argues that post–World War I biases of civil engineers led to a premature decline in wooden aircraft construction.

19. Charles Fayette Taylor, "Aircraft Propulsion: A Review of Aircraft Powerplants," Smithsonian Institution, *Annual Report* (1962), 269–74;

Charles L. Lawrance, "The Development of the Airplane Engine in the United States," International Civil Aeronautics Conference, *Papers* (Washington, D.C.: U.S. Government Printing Office, 1928) 411–21; Robert Schlaifer and Samuel D. Heron, *Development of Aircraft Engines and Fuels* (Cambridge: Harvard Graduate School of Business Administration, 1950), 156–98.

20. Bilstein, *Patterns*, 109–110, passim; Schlaifer and Heron, *Development*, 559–90.

21. Gwynn-Jones, *Farther and Faster*, 243–46. Rickenbacker quote from page 245.

22. Davies, *Airlines*, 115–16; Henry Ladd Smith, *Airways: The History of Commercial Aviation in the United States* (New York: Knopf, 1942), 133, 137, 160–70.

23. Robert F. van der Linden, *The Boeing 247: The First Modern Airliner* (Seattle: University of Washington Press, 1991), 25 and 226n6.

24. Miller and Sawers, *Technical Development*, 18, 50–54, 63; Peter Bowers, *Boeing Aircraft Since 1916* (London: Putnam, 1968), 175–78.

25. van der Linden, *Boeing 247*, 62–63; Mansfield, *Vision*, 98–100; Miller and Sawers, *Technical Development*, 68–69, 97–102.

26. van der Linden, *Boeing 247*, 29–32, 35–39.

27. Ibid., 53–58, 60.

28. Ibid., 1–2, 48, 51–52; Mansfield, *Vision*, 100–108.

29. Ibid., 67–69, 79–82; J. Parker Van Zandt, "Aviation Comes of Age," *Official World's Fair Weekly* (August 26, 1933): 5–7, copy in Century of Progress files, Chicago Historical Society.

30. David E. Lee, "Herbert Hoover and the Golden Age of Aviation," in William Leary, ed., *Aviation's Golden Age: Portraits from the 1920s and 1930s* (Iowa City: University of Iowa Press, 1989), 127–37, 147; Vander Meulen, *The Politics of Aircraft*, 1–8, 102–111, 203–06. Pilot insight courtesy of Tom Crouch, Air and Space Museum.

CHAPTER 3

1. On the Aeronautical Chamber of Commerce, see Jacob Vander Meulen, *The Politics of Aircraft: Building an American Military Industry* (Lawrence: University Press of Kansas), 188–91. The origins of the Institute of Aeronautical Sciences are summarized in its "Proceedings of the Founder's Meeting, Institute of the Aeronautical Sciences, Inc., held at Columbia University, New York, January 26, 1933," 6–10, 35–37 (copy in National Air and Space Museum files), and in Roger E. Bilstein, "Edward Pearson Warner and the New Air Age," from William M. Leary, ed., *Aviation's Golden Age*, 113–26. For a discussion of the origins of the Wings Club, see Clayton Knight, ed., *The History of the Wings Club: The First Twenty-five Years, 1942–1967* (New York: Wings Club, 1967).

2. For a useful summary of the light plane industry, see Tom Crouch, "General Aviation: the Search for a Market, 1910–1976," in *Two Hundred Years of Flight in America,* ed. Eugene M. Emme (San Diego: Univelt, 1977), 108–35. On Piper, see Devon Francis, *Mr. Piper and His Cubs* (Ames: Iowa State University Press, 1973), 3–73, 87; Walter J. Boyne, "Those Anonymous Cubs," *Aviation Quarterly* 1, no. 4 (Winter 1975): 252–80; "Two New Cubs," in Leighton Collins, ed., *Air Facts Reader, 1939–1941* (New York: Air Facts Press, 1974), 78–83; Bill Siuru, "The Piper Cub: Simplicity Takes Flight and Endures," *Mechanical Engineering* 112 (November 1990): 48–52.

3. Cessna Corporation has published a detailed corporate chronology, *An Eye to the Sky* (Wichita: Cessna, 1962). See also Mitch Mayborn and Bob Pickett, *Cessna Guidebook,* vol. I (Dallas: Flying Enterprise Publications, 1973); Edward H. Phillips, *Cessna: A Master's Expression* (Eagan, Minn.: Flying Books, 1985), passim; Peter Bowers, ed., *Yesterday's Wings* (Washington, D.C.: Aircraft Owners and Pilots Association, 1974), 20–21, 129; Paul Poberezny and S. H. Schmid, eds., *Wings of Memory* (Hales Corners, Wis.: Experimental Aircraft Association, 1969), 12–13, 19–21, drawn from contemporaneous articles in *Aero Digest.*

4. William H. McDaniel, *The History of Beech: Four Decades of Aeronautical and Aerospace Achievements* (Wichita: McCormick-Armstrong, 1976), 1–42; Joseph P. Juptner, *U.S. Civil Aircraft,* vol. 8 (Fallbrook, Calif.: Aero Publishers, 1980), 43–46, 53–55.

5. Phillips, *Cessna,* 113–14, 120–23; Welman Austin Shrader, *Fifty Years of Flight: A Chronicle of the Aviation Industry in America, 1903–1953* (Cleveland: Eaton Manufacturing Company, 1953), 54, passim. See also Frank J. Rome and Craig Miner, *Borne on the Wind: A Century of Kansas Aviation* (Wichita: Wichita Eagle and Beacon Pub. Co., 1994), which covers the light plane industry.

6. Ronald Miller and David Sawers, *The Technical Development of Modern Aviation* (New York: Praeger, 1970) covers the major technological changes, their innovators, and significance. On O-rings, see Walter Vincenti, "The Retractable Airplane Landing Gear and the Northrop 'Anomaly': Variation-Selection and the Shaping of Technology," *Technology and Culture* 35 (January 1994), 1–33. For the significance of MTOW, see Richard K. Smith, "The Intercontinental Airliner and the Essence of Airplane Performance," *Technology and Culture* 24 (July 1983): 428–449; Richard K. Smith, "The Weight Envelope: An Airplane's Fourth Dimension...Aviation's Bottom Line," *Weight Engineering* 47 (Summer 1987): 32–47. The latter includes pointed reminders about geographical factors, and my appreciation of this aspect was sharpened through conversations with R.E.G. Davies, National Air and Space Museum, during my residence there in 1992–93. For a European

acknowledgment of American advantage in terms of MTOW and the geographical element, see Roy Braybrook, "The Sixties, the Thirties, and the American Challenge," *Air International* 32 (March 1987): 129–34.

7. There are many sources that recount the legend of the DC-3 origins. For a readable, informative summary, see Frederick Allen, "The Letter that Changed the Way We Fly," *American Heritage of Invention and Technology*, vol. 4 (Fall 1988): 6–13. Douglas J. Ingells, *The Plane that Changed the World* (Fallbrook, Calif.: Aero, 1966), is a detailed narrative. See also Miller and Sawers, *Technical Development*, 98–107, and Richard Hallion, *Legacy of Flight: The Guggenheim Contribution to American Aviation* (Seattle: University of Washington Press, 1977), 192–98, which notes details of technical design and wind tunnel tests in facilities funded through the Guggenheim Foundation. Kenneth Munson, *Airliners between the Wars, 1919–1939* (New York: Macmillan, 1972), provides succinct information on the DC-2/3 evolution as well as numerous other designs of the era.

8. The MacRobertson event, well remembered in the lore of air racing, took on more significance during the late 1980s, when European pique about American leadership in global airliner sales sharpened controversy about technological leadership and its origins. A standard account of the race is contained in C.R. Roseberry, *The Challenging Skies: The Colorful Story of Aviation's Most Exciting Years, 1919–1939* (New York: Doubleday, 1969). My account here is largely based on Terry Gwynn-Jones, *Farther and Faster: Aviation's Adventuring Years, 1909–1939* (Washington, D.C.: Smithsonian Institution Press, 1991), 253–59.

9. R.E.G. Davies, *A History of the World's Airlines* (London: Oxford University Press, 1967), passim; Munson, *Airliners*, 76–77, 162–65 ff; Arthur Pearcy, *DC-3* (New York: Ballantine, 1975), passim; Miller and Sawers, *Technical Development*, 48–127; Vander Meulen, *Politics*, 99. On the aesthetics of aeronautical engineering, see Donald J. Bush, *The Streamlined Decade* (New York: George Braziller, 1975).

10. On the often neglected significance of technical advances in engineering and construction of flying boats, see Richard K. Smith, "The Intercontinental Airliner," 428–49.

11. For colorful, superbly illustrated surveys of air transportation and the dirigible era, see Oliver E. Allen, *The Airline Builders* (Alexandria, Va.: Time-Life Books, 1981); Douglas Botting, *The Giant Airships* (Alexandria, Va.: Time-Life Books, 1980).

12. Frank Cunningham, *Sky Master: The Story of Donald Douglas and the Douglas Aircraft Company* (Philadelphia: Dorrance, 1943), 219–25; Peter W. Brooks, *The Modern Airliner: Its Origins and Development* (London: Putnam, 1961), 91–111; Miller and Sawers, *Technical Development*, 129–41.

13. Ingells, *Plane*, 235–36; Cunningham, *Sky Master*, 267–68, 272; Miller and Sawers, *Technical Development*, 131–32, 141–42; Kenneth Hudson, *Air Travel: A Social History* (Totowa, N.J.: Rowman and Littlefield, 1972), 109. The complex issue of defining flying qualities and designing them into the airplane is described in Walter Vincenti, *What Engineers Know and How They Know It: Analytical Studies from Aeronautical History* (Baltimore: Johns Hopkins University Press, 1990), 51–111; for his discussion of flush riveting, 172–99.

14. Clive Irving, *Wide Body: The Triumph of the 747* (New York: Morrow, 1993), 38–48.

15. M. J. Hardy, *The Lockheed Constellation* (New York: Arco, 1973), 10–18, 61.

16. Michael H. Gorn, *The Universal Man: Theodore von Karman's Life in Aeronautics* (Washington, D.C.: Smithsonian Institution Press, 1992), especially pages 55–110.

17. James R. Hansen, *Engineer in Charge: A History of the Langley Aeronautical Laboratory, 1917–1958* (Washington, D.C.: U.S. Government Printing Office, 1987), 109–18, 174, passim; Alex Roland, *Model Research: The National Advisory Committee for Aeronautics, 1915–1958*, vol. I (Washington, D.C.: U.S. Government Printing Office, 1985), 119–20, 127–28, 235–36; David Anderton, *Sixty Years of Aeronautical Research, 1917–1977* (Washington, D.C.: U.S. Government Printing Office, 1978), 16–19.

18. Roger Bilstein, *Orders of Magnitude: A History of the NACA and NASA, 1915–1990* (Washington, D.C.: U.S. Government Printing Office, 1989), 21–22, passim.

19. Virginia P. Dawson, *Engines and Innovation: Lewis Laboratory and American Propulsion Technology* (Washington, D.C.: U.S. Government Printing Office, 1991); Hansen, *Engineer in Charge*, 148–49, 158–61, 194–202; Roland, *Model Research*, 162–64, 173, 177–98.

20. The negative role of congressional interference is the theme of Vander Meulen, *Politics*. Exports in particular are noted in Vander Meulen, 109–11; John B. Rae, *Climb to Greatness: The American Aircraft Industry, 1920–1960* (Cambridge: MIT Press, 1968), 77–78; and Anthony Sampson, *The Arms Bazaar: From Lebanon to Lockheed* (New York: Viking Press, 1977), 76–78. Wayne Biddle, *Barons of the Sky: From Early Flight to Strategic Warfare* (New York: Simon and Schuster, 1991), also includes commentary on exports by Martin, Douglas, Northrop, and others.

21. Ray Wagner, *American Combat Planes* (New York: Doubleday, 1982), 368–80; Bill Gunston, *Grumman: Sixty Years of Excellence* (New York: Orion, 1988), 24–37, 42–47.

22. For a convenient survey on the Army's aviation trends, with good illustrations and insights, see David Anderton, *The History of the U.S. Air*

Force (New York: Crescent Books, 1981), 31–42. Other convenient references include Andrew W. Waters, *All the U.S. Air Force Airplanes, 1907–1983* (New York: Hippocrene Books, 1983), 80–90, 243–53; Wagner, *American Combat Planes*, 193–207, 232–51, passim. Recent books with pertinent background include Stephen L. McFarland and Wesley P. Newton, *To Command the Sky: The Battle for Air Superiority over Germany, 1942–1944* (Washington, D.C.: Smithsonian Institution Press, 1991), 11–39.

23. My discussion of the bomber issue profited considerably from Robert van der Linden, "The Struggle for the Long-Range Heavy Bomber: The United States Army Air Corps, 1934–1939" (M.A. thesis, University of Denver, 1981), 8–9, 28–31, 141; see also Harold Mansfield, *Vision: A Saga of the Sky* (New York: Duell, Sloan, and Pearce, 1956), 114; Wesley Frank Craven and James Lea Cate, eds., *Men and Planes* (Washington, D.C.: U.S. Government Printing Office, 1955), 202–04; Henry H. Arnold, *Global Mission* (New York: Harper, 1942), 179–80; Michael S. Sherry, *The Rise of Armageddon* (New Haven: Yale University Press, 1987), 1–78.

24. Thomas M. Coffey, *Hap: The Story of the U.S. Air Force and the Man Who Built It* (New York: Viking, 1982), 171–98; Vander Meulen, *Politics*, 1–7, 117–46, 193–220.

25. Craven and Cate, *Men*, 191–301, 301, 313; Rae, *Climb*, 110, 115, 127; William Glenn Cunningham, *The Aircraft Industry: A Study in Industrial Location* (Los Angeles: Lorrin L. Morrison, 1951), 77.

26. William Hess, *Fighting Mustang: The Chronicle of the P-51* (Garden City, N.Y.: Doubleday, 1970), 3; Jeffrey Ethel, *Mustang: A Documentary History* (London: Jane's, 1981), 9–10; David Birch, *Rolls-Royce and the Mustang* (Derby, U.K.: Rolls-Royce Heritage Trust, 1987), 10, 145–46.

27. Ethell, *Mustang*, 10–13; Waters, *U.S. Air Force*, 263–65; Ronald Harker, *The Engines Were Rolls-Royce* (New York: Macmillan, 1979), 68–71.

28. David Boulton, *The Grease Machine* (New York: Harper and Row, 1978), 28–30.

29. Rae, *Climb*, 113–18, 123, 143–45; Alfred Goldberg, ed., *A History of the United States Air Force, 1907–1957* (Washington, D.C.: U.S. Government Printing Office, 1957), 91; Vander Meulen, *Politics*, 147–81, 207–17, 222.

30. Russell E. Lee, "The Impact of Victory through Air Power," part I, *Air Power History* (Summer 1993): 3–4, passim; part II (Fall 1993): 28–30.

31. Ibid., part I (Summer 1993): 5, 7–13; part II (Fall 1993): 20–30.

32. Goldberg, *History of USAF*, 91; Juanita Loveless, ed., *Rosie the Riveter* (Long Beach: California State College, 1983), passim, especially pages 259–70; Herbert R. Northrup, *The Negro in the Aerospace Industry*,

Racial Policies of American Industry, Report no. 2 (Philadelphia: Wharton School of Finance and Commerce, 1968), passim.

33. Irving Brinton Holley, Jr., *Buying Aircraft: Materiel Procurement for the Army Air Forces* (Washington, D.C.: U.S. Government Printing Office, 1964), 186–93, 539–47.

34. E.C. Barksdale, *The Genesis of the Aviation Industry in North Texas* (Austin: University of Texas, Bureau of Business Research, 1958), 1–25.

35. Holley, *Buying Aircraft*, 518–38, Appendix C, p. 580.

36. Eugene E. Bauer, *Boeing in Peace and War* (Enumclaw, Washington: TABA Publishing, 1991), 137–38; Thomas F. Collison, *The Superfortress Is Born* (New York: Duell, Sloan, and Pearce, 1945), 3.

37. Collison, *Superfortress*, passim; David Anderton, *B-29 Superfortress at War* (New York: Scribner's, 1978), 11–14, 24–25; Kenneth Wheeler, *Bombers over Japan* (Alexandria, Va.: Time-Life Books, 1982), passim; Mansfield, *Vision*, 167–73; 178–81; Jacob Vander Meulen, "Who Built the B-29" (unpublished ms., lent by the author, 1993), passim. Dr. Vander Meulen generously shared this latter study with me during my fellowship year at the National Air and Space Museum, 1992–93.

38. Collison, *Superfortress*, 161–65.

39. Rae, *Climb*, 121–22, 155–57.

40. For detailed summaries and assessments of the production record, including individual manufacturers, see Holley, *Buying Aircraft*, 548–87. For an interpretive comparison of the U.S. record with other nations, see Jonathan Zeitlin, "Flexibility and Mass Production at War: Aircraft Manufacture in Britain, the United States, and Germany, 1939–1945," *Technology and Culture* 36 (January 1995): 46–79.

CHAPTER 4

1. Paul E. Ceruzzi, *Beyond the Limits: Flight Enters the Computer Age* (Cambridge: Cambridge University Press, 1989), 44–45; Fred Kaplan, *The Wizards of Armageddon* (New York: Simon & Schuster, 1983), 58–64; Jacob Neufeld, ed., *Reflections on Research and Development in the United States Air Force* (Washington, D.C.: U.S. Government Printing Office, 1992), 1–15, passim; Alfred Goldberg, ed., *History of the United States Air Force, 1907–1957* (New York: Van Nostrand, 1957), 99–120, 197–226, offers a useful summary.

2. Donald J. Mrozek, "The Truman Administration and the Enlistment of the Aviation Industry in Postwar Defense," *Business History Review* 48 (Spring 1974): 73–94; Bill Gunston, *World Encyclopedia of Aircraft Manufacturers* (Annapolis, Md.: Naval Institute Press, 1989), passim; Robert W. Fausel, *Whatever Happened to Curtiss-Wright* (Manhattan, Kans.: Sunflower Press, 1990). Fausel himself had been a Curtiss-Wright

executive, and his memoir includes comments from numerous correspondents who had been with the company.

3. E. E. Bauer, *Boeing in Peace and War* (Enumclaw, Wash.: TABA Publishing, 1991), 153–63.

4. Simonson, *Aircraft Industry*, 227–41; Aerospace Industries Association, "Seventy-Fifth Anniversary Perspectives," (Washington, D.C.: AIA, 1994), brochure in author's collection.

5. James R. Hansen, *Engineer in Charge: A History of the Langley Aeronautical Laboratory, 1917–1958* (Washington, D.C.: U.S. Government Printing Office, 1987), Chapter 8, esp. pp. 221–41; Ronald Miller and David Sawers, *The Technical Development of Modern Aviation* (New York: Praeger, 1970), 162–63.

6. Richard P. Hallion, "Lippisch, Gluhareff and Jones: The Emergence of the Delta Platform and the Origins of the Swept Wing in the United States, *Aerospace Historian* 26 (March 1979): 1–10; Hansen, *Engineer in Charge*, 279–90.

7. Miller and Sawers, *Technical Development*, passim; Hansen, *Engineer in Charge*, 243; Roland, Model Research, 194, 203–19.

8. Clarence Lasby, *Project Paperclip: German Scientists and the Cold War* (New York: Atheneum, 1975); Dawson, *Engines and Innovation*, 65–72, 78–101, 141–44; Bill Gunston, *World Encyclopedia of Aero Engines* (London: Patrick Stephens, 1989), 9–11, 61–63, 119–21, 185–87; Wagner, *American Combat Planes*, 443, 505–7.

9. Ceruzzi, *Beyond the Limits*, 12–37, 63–77, 83, 115–21; Herman O. Stekler, *The Structure and Performance of the Aerospace Industry* (Berkeley: University of California Press, 1965), especially Chapter two. Shrinking computer circuitry and architecture contributed to home computers, pocket calculators, and similar products.

10. Robin Higham, *Air Power: A Concise History* (New York: St. Martin's Press, 1972), offers a comprehensive summary. For details as well as useful commentary on a wide variety of designs, see Kenneth Munson, *Fighters in Service: Attack and Training Aircraft since 1960* (New York: Macmillan, 1975), and Wagner, *American Combat Planes*. For a concise critique of the Korean and Vietnamese experience and reassessments, see Richard P. Hallion, *Storm over Iraq: Air Power and the Gulf War* (Washington, D.C.: Smithsonian Institution Press, 1992), especially 1–39.

11. For an informative and superbly illustrated survey, see Warren J. Young, *The Helicopters* (Alexandria, Va.: Time-Life Books, 1982). Military aspects are covered in Bill Gunston and John Batchelor, *Helicopters at War* (New York: Chartwell Books, 1977), and David W. Wragg, *Helicopter at War: A Pictorial History* (New York: St. Martin's, 1983).

12. Wagner, *American Combat Planes*, passim; Clive Irving, *Wide Body: The Triumph of the 747* (New York: Morrow, 1993), 84, 86–94, 165,

196–99; Bill Gunston, *Bombers of the West* (New York: Scribner's, 1973), 126–53; John E. Steiner, "Jet Aviation Development: A Company Perspective," in Boyne and Lopez, *The Jet Age*, 141–48.

13. Jay Miller, *Convair B-58* (Arlington, Tex.: Aerofax, 1985), 17–27, 35–55; Gunston, *Bombers*, 185–212; Cargill Hall, "The B-58 Bomber: Requiem for a Welterweight," *Air University Review* 33 (November–December 1981): 44–56.

14. Harold Mansfield, *Vision: A Saga of the Sky* (New York: Duell, Sloan and Pearce, 1956), 175–82.

15. Gunston, *Bombers*, 185–212; Jeff Ethell and Joe Christy, *B-52 Stratofortress* (New York: Scribner's, 1981), passim. Martin Streetly, *World Electronic Warfare Aircraft* (London: Jane's, 1983), 61–63; Kenneth Schaffel, *The Emerging Shield: The Evolution and Development of USAF Forces for Continental Air Defense to 1960* (Washington, D.C.: U.S. Government Printing Office, 1989), passim; Kenneth Munson, *Bombers in Service: Patrol and Transport Aircraft since 1960* (New York: Macmillan, 1975), passim.

16. Bill Gunston, *Early Supersonic Fighters of the West* (New York: Scribner's, 1976), 171–73.

17. Richard P. Hallion, *On the Frontier: Flight Research at Dryden, 1946–1981* (Washington, D.C.: U.S. Government Printing Office, 1984), details military factors of flight research and influences on production aircraft. Aspects of tooling are noted in Charles Bright, *The Jet Makers: The Aerospace Industry from 1945 to 1972* (Lawrence, Kans.: Regents Press, 1978), 119–25. Examples of many contemporaneous manufacturing techniques can be found in Frank Wilson and Walter R. Prange, eds., *Tooling for Aircraft and Missile Manufacture* (New York: McGraw-Hill, 1964), 2, 109–11, 302–13. Other commentary from Gunston, *Early Fighters*, 143–46; Miller and Sawers, *Technical Development*, 175–77.

18. Bright, *Jet Makers*, 131–47.

19. Ibid., 109–13, 149–67. See also Herman O. Stekler, *The Structure and Performance of the Aerospace Industry* (Berkeley: University of California Press, 1965); and Edward H. Heinemann and Rosario Rausa, *Ed Heinemann: Combat Aircraft Designer* (Annapolis, Md.: Naval Institute Press, 1980).

20. Arthur Reed, *F-104 Starfighter* (New York: Scribners, 1981), passim; Douglas J. Ingells, *L-1011 TriStar and the Lockheed Story* (Fallbrook, Calif.: Aero Publishers, 1973), passim.

21. Reed, *F-104 Starfighter*, 6, 8–10, 18–21; Ingells, *TriStar*, 137–39.

22. Reed, *F-104 Starfighter*, 39–41, 62, 68, 71; Gunston, *Early Fighters*, 184–214.

23. Kenneth Munson, *Fighters, Attack, and Training Aircraft* (New York: Macmillan, 1966), 98–99; Wagner, *Combat Planes*, 469, note 5; David

Boulton, *The Grease Machine* (New York: Harper and Row, 1978), 85–110, 158–84; Harker, *Rolls-Royce*, 138.

24. Randolph P. Kucera, *The Aerospace Industry and the Military: Structural and Political Relationships* (Beverly Hills, Calif.: SAGE Publications, series 03-023, 1974), 44–45.

25. Carroll W. Purcell, ed., *The Military Industrial Complex* (New York: Harper and Row, 1972), includes informed analysis as well as Eisenhower's famous remarks, 204–8.

26. Robert F. Coulam, *Illusions of Choice: The F-111 and the Problem of Weapons Acquisition Reform* (Princeton: Princeton University Press, 1977), 203–15; Bill Gunston, *F-111* (New York: Arco, 1983), 7, passim. Coulam's study gave low marks to concurrency's supposed economy.

27. Kucera, *Aerospace Industry*, 45–55; Coulam, *Illusions of Choice*, 203–15, Gunston, *F-111*, 7–9.

28. Coulam, *Illusions of Choice*, 77–80, and extended footnotes, nos. 95–100.

29. Gunston, *F-111*, 33–39.

30. Ron Westrum and Howard Wilcox, "Sidewinder," *Invention and Technology* 5 (Fall 1989): 57–59.

31. Ibid., 59–63; Aerospace Industries Association, *The 1969 Aerospace Yearbook* (Washington, D.C.: Books, Incorporated, 1969), R-143; Air Force Association, *Air Force Almanac* (May 1993): 145–46; Christopher Chant, *World Encyclopedia of Modern Air Weapons* (Wellingborough, England: Patrick Stephens Limited, 1988), 212–13, 215–18; John C. McLaurin, *Accelerated Production: The Air-to-Air Missile Case* (Washington, D.C.: National Defense University Press, 1981), 57–61 and passim.

32. John Edward Wilz, *Democracy Unchallenged: The United States since World War II* (New York: Harper and Row, 1990), 243–45.

CHAPTER 5

1. Constance Green and Milton Lomask, *Vanguard: A History* (Washington, D.C.: Smithsonian Institution Press, 1971); Milton W. Rosen, "October 4, 1957: Sputnik Beeps, America Responds," *Astronautics and Aeronautics* (October 1977): 20–23.

2. For a survey of the Eisenhower administration's situation, see Robert A. Divine, *The Sputnik Challenge: Eisenhower's Response to the Soviet Satellite* (New York: Oxford University Press, 1993), and especially, R. Cargill Hall, "The Eisenhower Administration and the Cold War: Framing American Astronautics to Serve National Security," *Prologue* 27 (Spring 1995), 59–72.

3. See, for example, Loyd S. Swenson, Jr., James M. Grimwood, and Charles C. Alexander, *This New Ocean: A History of Project Mercury*

(Washington, D.C.: U.S. Government Printing Office, 1966); Walter McDougal, *The Heavens and the Earth: A Political History of the Space Age* (New York: Basic Books, 1985).

4. Wernher von Braun and Fred I. Ordway III, *History of Rocketry and Space Travel* (New York: Crowell, 1975), is still an informative and useful work. Frank Winter, *Rockets into Space* (Cambridge: Harvard University Press, 1990), presents a concise, accurate summary of booster developments; Frank Winter, *Prelude to the Space Age: The Rocket Societies, 1924–1940* (Washington, D.C.: Smithsonian Institution Press, 1983), 14, 75–78, is an important study.

5. Winter, *Prelude*, 73–85; Andrew G. Haley, *Rocketry and Space Exploration* (New York: Van Nostrand, 1958), 57–58. The latter is an especially useful memoir by one who became a central figure in American rocket societies as early as the 1930s.

6. Frank Malina, "Origins and First Decade of the Jet Propulsion Laboratory," in Eugene M. Emme, ed., *The History of Rocket Technology: Essays on Research, Development, and Utility* (Detroit: Wayne State University Press, 1964), 52, 54.

7. Haley, *Rocketry*, 69; von Braun and Ordway, *Rocketry and Space Travel*, 81–85, 93.

8. Haley, *Rocketry*, 71–91, 93–104; Clayton R. Koppes, *JPL and the American Space Program* (New Haven: Yale University Press, 1982), 9–17; von Braun and Ordway, *Rocketry and Space Travel*, 93–104.

9. Frederick Ordway and Frank Winter, "Pioneering Commercial Rocketry . . . Reaction Motors . . . Corporate History," *Journal of the British Interplanetary Society* 36 (December 1983): 543; Frank Winter and Frederick Ordway, "Pioneering Commercial Rocketry . . . Reaction Motors . . . Projects," Ibid. 38 (April 1985): 158–59.

10. Haley, *Rocketry*, 156–59; Ordway and Winter, "Pioneering . . . Reaction Motors . . . Corporate," 545–46, 549; Winter and Ordway, "Pioneering . . . Reaction Motors . . . Projects," 164.

11. Haley, *Rocketry*, 156–90; Frederick Ordway and Ronald C. Wakeford, *International Missile and Spacecraft Guide* (New York: McGraw-Hill, 1960), passim; J.L. Atwood, *North American Rockwell: Storehouse of High Technology* (New York: Newcomen Society, 1970); miscellaneous corporate brochures in author's files.

12. Roger E. Bilstein, *Stages to Saturn: A Technological History of the Saturn Launch Vehicles* (Washington, D.C.: U.S. Government Printing Office, 1980), 13–18; Frederick Ordway and Mitchell Sharpe, *The Rocket Team: From the V-2 to the Saturn Moon Rocket* (New York: Crowell, 1979); Ordway and Wakeford, *International Guide*, passim.

13. Ernest G. Schwiebert, "USAF's Ballistic Missiles, 1954–1964," *Air Force and Space Digest* 47 (May 1964): 56–69, 71–75, 78–83, 87–90;

Kenneth Werrel, *The Evolution of the Cruise Missile* (Maxwell Air Force Base, Ala.: Air University Press, 1985), passim; Jacob Neufeld, *The Development of Ballistic Missiles in the United States Air Force, 1945–1960* (Washington, D.C.: Office of Air Force History, 1990), passim; Michael H. Gorn, *The Universal Man: Theodore von Karman's Life in Aeronautics* (Washington, D.C.: Smithsonian Institution Press, 1992), 97–117; R. Cargill Hall, correspondence to the author, September 29, 1993.

14. Schwiebert, "USAF's Ballistic Missiles," 86–89; J.S. Butz, "The USAF Missile Program: A Triumph of Orderly Engineering," *Air Force and Space Digest* 47 (May 1964): 182–83; Neufeld, *Development of Ballistic Missiles*, 211–12, 126–28, 180; Robert Sheehan, "Thompson Ramo Wooldridge: Two Wings in Space," *Fortune* 57 (February 1963): 95–99.

15. Neufeld, *Development of Ballistic Missiles*, 223–32; Ordway and Wakeford, *International Guide*, passim; Kenneth W. Gatland, *Missiles and Rockets*, (New York: Macmillan, 1975), 130–37.

16. Neufeld, *Development of Ballistic Missiles*, 122–31, 213–14, 227–32, 237; Gatland, *Missiles and Rockets* 132–37; Aerospace Industries Association, *The 1969 Aerospace Yearbook* (Washington, D.C.: Books, Incorporated, 1969), R-121.

17. Neufeld, *Development of Ballistic Missiles*, 239–42, 244, 163n1; Schwiebert, "USAF's Ballistic Missiles," 93, 139; Gatland, *Missiles and Rockets*, passim. Based on the special issue of *Air Force and Space Digest* of May 1964, Schwiebert edited and published *A History of the United States Air Force Ballistic Missiles* (New York: Praeger, 1965). My citations come from the original periodical version, which includes useful illustrations, advertisements and miscellaneous contemporaneous information.

18. Gatland, *Missiles and Rockets*, 154–60; Aerospace Industries Association, *1969 Yearbook*, R-124. James Baar and William E. Howard, *Polaris!* (New York: Harcourt Brace, 1960), represents a journalistic, popularized account. For a detailed philosophical analysis of Polaris/Poseidon/Trident missile technology and guidance systems, see Donald MacKenzie, *Inventing Accuracy: A Historical Sociology of Nuclear Missile Guidance* (Cambridge: MIT Press, 1990).

19. Harvey Sapolsky, *The Polaris System Development: Bureaucratic and Programmatic Success in Government* (Cambridge: Harvard University Press, 1972), 44–48, 110–12, 114n56, 128–30, 230–54; Roger E. Bilstein, *Stages to Saturn*, 283–91.

20. Randy Lieberman, "The *Colliers* and Disney Series," in Frederick Ordway et al., *Blueprint for Space: Science Fiction to Science Fact* (Washington: Smithsonian Institution Press, 1992), 135–46.

21. Loyd Swenson, James Grimwood, and Charles C. Alexander, *This New Ocean: A History of Project Mercury* (Washington, D.C.: U.S. Government Printing Office, 1966), passim; McDougal, *The Heavens and the*

Earth, passim. On the lunar landing decision in particular, see John Logsdon, *The Decision to Go to the Moon: Project Apollo and the National Interest* (Cambridge: MIT Press, 1970). Logsdon also wrote an insightful update on the Apollo decision, "The Challenge of Space: Linking Aspirations and Political Will," in Ordway et al., *Blueprint,* 147–52. Aspects of the "fertile crescent" are noted in Bilstein, *Stages to Saturn,* 390–94.

22. Courtney G. Brooks, James M. Grimwood, and Loyd S. Swenson, Jr., *Chariots for Apollo: A History of Manned Lunar Spacecraft* (Washington, D.C.: U.S. Government Printing Office, 1979). The budget summary is extrapolated from pp. 409–11. On AVRO and the Canadians, see Henry Dethloff, *Suddenly, Tomorrow Came . . . A History of the Johnson Space Center* (Washington, D.C.: U.S. Government Printing Office, 1994), 20–26.

23. Bilstein, *Stages to Saturn,* 422; James E. Webb, "A Perspective on Apollo," in Edgar M. Cortright, *Apollo Expeditions to the Moon* (Washington, D.C.: U.S. Government Printing Office, 1975), 12, 17; Robert R. Gilruth, "I Believe We Should Go to the Moon," in Cortright, *Apollo Expeditions,* 19–34.

24. Brooks et al., *Chariots,* 143–66, 196–201, 213–45, 399; see also Richard S. Lewis, *The Voyages of Apollo: The Exploration of the Moon* (New York: Quadrangle, 1974).

25. Bilstein, *Stages to Saturn,* 29–48, 74–83, 137–90, 351–55. For von Braun's own recollections, see his essay, "Saturn the Giant," in Cortright, ed., *Apollo Expeditions,* 41–58. An excellent summary of rocket technology, including missile developments and relevance to the American space program, is Frank Winter, *Rockets into Space* (Cambridge: Harvard University Press, 1990).

26. Bilstein, *Stages to Saturn,* passim; pp. 192–233 on the S-IC and S-II stages, pp. 261–92 on management.

27. Ibid., 347–72.

CHAPTER 6

1. Aerospace Industries Association, *Aerospace Facts and Figures: 1965* (Fallbrook, Calif.: Aero Publishers, 1966), 95; Aerospace Industries Association, *Aerospace Facts and Figures: 1975/76* (Washington, D.C.: McGraw-Hill, 1977), 72.

2. Theodore Paul Wright, "Britain's Influence in World Civilization: Past, Present, and Future," *Royal Aeronautical Society Journal, Proceedings,* (1966): 74–75.

3. Ronald Miller and David Sawers, *The Technical Development of Modern Aviation* (New York: Praeger, 1970), 288–89.

4. Bauer, *Boeing in Peace and War,* 150–52; Douglas J. Ingells, *747: Story of the Boeing Super Jet* (Fallbrook, Calif.: Aero Publishers, 1970), 85–102; Joseph Juptner, *U.S. Civil Aircraft,* vol. 9 (Fallbrook, Calif.: 1981), 45–48.

5. Douglas J. Ingells, *The McDonnell Douglas Story* (Fallbrook, Calif.: Aero Publishers, 1979), 122–27; M.J. Hardy, *Lockheed Constellation* (New York: Arco, 1973), passim; Juptner, *U.S. Civil Aircraft,* vol. 8, 279–82; Kenneth Munson, *Airliners since 1946* (New York: Macmillan, 1975), 124–26, 129–31; U.S. Civil Aeronautics Board, *Handbook of Airline Statistics* (Washington, D.C.: U.S. Government Printing Office, 1962), 452–55, 461, 530.

6. Munson, *Airliners since 1946,* 118–27; David Mondey, compiler, *Encyclopedia of the World's Commercial and Private Aircraft* (New York: Crescent, 1981), 112–13, 156–57, 196.

7. Kenneth Munson, *Bombers in Service: Patrol and Transport Aircraft Since 1960* (New York: Macmillan, 1975), 121–22.

8. See, for example, *Life* 19 (August 13, 1945): 48.

9. *Holiday* 7 (August 1947): 79; *Holiday* 12 (October 1952): 2; Ibid. (November 1952): 1, 25, 92, 149; Ibid. (December 1952): 40, 164; *National Geographic* (December 1965): 10.

10. Bauer, *Boeing,* 128–37; Miller and Sawers, *Technical Development,* 177–90.

11. Bauer, *Boeing,* 186–87; Clive Irving, *Wide Body: The Triumph of the 747* (New York: Morrow, 1993), 127–28; Martin Caidin, *Boeing 707* (New York: Ballantine, 1959).

12. Walter J. Boyne and Donald Lopez, eds., *The Jet Age: Fifty Years of Jet Aviation* (Washington, D.C.: Smithsonian Institution Press, 1979), 152–57.

13. Miller and Sawers, *Technical Development,* 193–95.

14. Ibid., 196–203, 279; Munson, *Airliners,* 148–49; Bauer, *Boeing,* 293–94; John Newhouse, *The Sporty Game* (New York: Knopf, 1983), passim.

15. Bauer, *Boeing,* 219–25; Miller and Sawers, *Technical Development,* 199–200; Mondey, *Commercial and Private Aircraft,* 56–62; NASA, *Progress in Aircraft Design Since 1903* (Washington, D.C.: U.S. Government Printing Office, 1974), 82.

16. Miller and Sawers, *Technical Development,* 266–76; Bauer, *Boeing,* 226–28; Steiner, "Jet Aviation," 158–62; Bright, *Jet Makers,* 116; Harold Mansfield, *Billion Dollar Battle: The Story behind the "Impossible" 727 Project* (New York: David McKay, 1965), passim.

17. NASA, *Progress,* 86; Ingells, *McDonnell Douglas,* 137–42; Munson, *Airliners,* 150–51; Mondey, *Commercial and Private Aircraft,* 139–43.

18. Irving, *Wide Body*, 172–74.

19. Frank Kingston Smith, "An Appreciation of the Social, Economic, and Political Issues of General Aviation," Report GA-300-133, (Washington, D.C.: Department of Transportation, Federal Aviation Administration, June 1977), passim; General Aviation Manufacturers Association, "The General Aviation Story" (Washington, D.C.: GAMA, 1977), 2–4, 7, 9; GAMA, "1977 Statistical Data" (Washington, D.C.: GAMA, 1977), 8–9. Copies of all above in author's collection.

20. R. Dixon Speas Associates, *The Magnitude and Economic Impact of General Aviation, 1968–1980* (Manhasset, N.Y.: Aero House, 1970), 13, passim; Aerospace Industries Association, *Aerospace Facts and Figures, 1975/76*, 82; General Aviation Manufacturers Association, "The General Aviation Story," passim. Aspects of influence of U.S. light plane products on additional sales evolved from figures cited and from anecdotal commentary during interviews for an oral history with 15 senior executives from Beech, Cessna, and Learjet during the spring of 1978. Notes in author's files.

21. John Paul Andrews, *Your Personal Plane* (New York: Duell, Sloan and Pearce, 1945), v–viii, 65.

22. Piper ad from *Life* 19 (August 13, 1945), 14; Frank Kingston Smith, "The Turbulent Decade," in *Flying: 50th Anniversary Issue* (September 1977): 203–5; "New Planes for Popular Flying," *Fortune* 33 (February 1946): 124–29.

23. Juptner, *U.S. Civil Aircraft*, vol. 8, 241–44, 283–87; Dorr Carpenter and Mitch Mayborn, *Ryan Guidebook* (Dallas: Flying Enterprise, 1975), 70–71.

24. The fortunes of many postwar hopefuls can be followed in Shrader, *Fifty Years of Flight*, 88–174, covering the years 1945–1953. For roadable hopefuls, see James R. Chiles, "Flying Cars Were a Dream that Never Got Off the Ground," *Smithsonian* 19 (February 1989): 144–62. The significance of careful product improvement, dealers, and market niche is implicit in such books as Joe Christy, *The Single-Engine Cessnas* (New York: Sports Car Press, 1971), and William H. McDaniel, *The History of Beech: Four Decades of Aeronautical and Aerospace Achievement* (Wichita: McCormick-Armstrong, 1976).

25. Smith, "The Turbulent Decade," 205–8. For an informative survey of the light plane industry in Kansas, see Frank J. Rowe and Craig Miner, *Borne on the South Wind: A Century of Kansas Aviation* (Wichita: Wichita Eagle and Beacon Publishing Co., 1994).

26. Devon Francis, *Mr. Piper and His Cubs* (Ames: Iowa State University Press, 1973), 119–81.

27. Cessna Aircraft Company, *An Eye to the Sky* (Wichita: Cessna, 1962), passim, is a detailed chronology issued by the company. See also

Edward H. Phillips, *Cessna: A Master's Expression* (Eagan, Minn.: Flying Books, 1985), passim.

28. McDaniel, *History of Beech*, 65–88; Larry A. Ball, *Those Incomparable Bonanzas* (Wichita: McCormick-Armstrong, 1971), 1–48; Ralph Harmon, the former Bonanza project manager, interview with author (Houston, Tex.: January 7, 1975). The fortunes of Cessna, Beech, and several other postwar aspirants are covered in Rowe and Minor, *Borne on the Wind*.

29. Smith, "Turbulent Decade," 207–8; Max Karant, "The Underdog," in *Flying: 50th Anniversary Issue* (September 1977): 206; Kenneth Munson, *Private Aircraft: Business and General Purpose since 1946* (New York: Macmillan, 1967), 7–13, passim, is an excellent survey; Wayne Thoms, *Flying for Fun or Business* (New York: Arco, 1967). The last gives a good contemporaneous feel for the evolution of different aircraft types and customer choices.

30. Munson, *Private Aircraft*, 146–47; Thoms, *Flying*, 104–5, 107–8; Shrader, *Fifty Years of Flying*, passim, all include examples of such conversions and modified aircraft. For a brief commentary on these iconoclastic firms and their imaginative founders, see Jerald A. Slocum, "The Modifiers," in *Flying: 50th Anniversary Issue* (September 1977): 259.

31. From advertisements reproduced in Phillips, *Wings of Cessna*, 31, 33, 36.

32. Piper trends are covered in Francis, *Mr. Piper*, 164–79, passim, and Munson, *Private Aircraft*, 121–23, 151–53. On Mooney, see Juptner, *U.S. Civil Aircraft*, vol. 9, pp. 19–20; Munson, *Private Aircraft*, 125–26.

33. Bill Gunston, *Grumman: Sixty Years of Excellence* (New York: Orion Books, 1988), 81, 84.

34. Munson, *Private Aircraft*, 169–70, Mondey, *Commercial and Private Aircraft*, 187, 222–23.

35. Mondey, 165–67; Robert J. Serling, *Little Giant: The Story of Gates Learjet* (Potomac, Md.: R.J. Serling, 1974); NASA, *Progress in Aircraft Design*, 84; James Greenwood, Gates Learjet public relations, interview with author (Washington, D.C.: March 31, 1978); Don Grommesh, Gates Learjet project engineer, interview with author (Wichita: March 7, 1978).

36. Fred E. Weick and James R. Hansen, *From the Ground Up: The Autobiography of an Aeronautical Engineer* (Washington, D.C.: Smithsonian Institution Press, 1988), 252–84, 307–20, 327; Francis, *Mr. Piper*, 186–87. Material on Leland Snow is based on miscellaneous clippings supplied by Snow and his marketing agency as well as telephone interviews with Snow conducted during 1984–85. These sources all evolved from the author's role as co-curator for the exhibit, "Reach for the Sky" (Institute of Texas Cultures, San Antonio, 1985–86) and co-author, with Jay Miller,

Aviation in Texas (Austin: Texas Monthly Press, 1985). All notes and clippings in author's collection.

37. Igor Sikorsky, *The Story of the Winged-S: Late Developments and Recent Photographs of the Helicopter, An Autobiography* (New York: Dodd, Mead, 1958); Walter Boyne and Donald S. Lopez, *Vertical Flight: The Age of the Helicopter* (Washington, D.C.: Smithsonian Institution Press, 1984); Mondey, *Commercial and Private Aircraft*, 48–50, 230; Bell Aircraft, miscellaneous brochures in author's collection.

CHAPTER 7

1. Paul Ceruzzi, *Beyond the Limits: Flight Enters the Computer Age* (Cambridge: MIT Press, 1989), 92–94.
2. Christy Campbell, ed., *Understanding Military Technology* (New York: Gallery Books, 1986), 66, 70, 143; Susan H.H. Young, "Gallery of Air Force Weapons," *Air Force Magazine* 76 (May 1993): 145.
3. Campbell, *Military Technology*, passim; Kenneth Gatland, *Missiles and Rockets* (New York: Macmillan, 1975), passim; Department of Defense, *Report of the Secretary of Defense to the President and the Congress, January 1993* (Washington, D.C.: U.S. Government Printing Office, 1993), passim.
4. NASA's own history program includes several titles on such major programs as ASTP and Skylab; summaries of these and other programs appear in Jane D'Alelio et al., NASA, *The First 25 Years, 1958–1983* (Washington, D.C.: U.S. Government Printing Office, 1983), and in Roger Bilstein and Frank Anderson, *Orders of Magnitude: A History of the NACA and NASA, 1915–1990* (Washington, D.C.: U.S. Government Printing Office, 1989). For a critical assessment of the *Challenger* shuttle explosion and of the agency itself, see Joseph J. Trento, *Prescription for Disaster: From the Glory of Apollo to the Betrayal of the Shuttle* (New York: Crown, 1987).
5. For a sweeping survey of the postwar European aerospace scene, see Walter A. McDougall, *The Heavens and the Earth*, passim. Europe's development of the supersonic transport is analyzed in Charles Burnet, *Three Centuries to Concorde* (London, United Kingdom: Mechanical Engineering Publications, 1979). This section of the book also rests on Roger Bilstein, "Evolution of the European Aerospace Industry," an unpublished report prepared as a NASA/American Society for Engineering Education Faculty Fellow (Johnson Space Center, History Office, 1986). See also Bernard Udis, "European Perspectives on International Aerospace," and Rae Angus, "The Tornado Project," in Martin Edmonds, *International Arms Procurement: New Directions* (New York: Pergamon Press,

1981), 115, 168–71, 176–79, 183; "Partnership in Major Engineering Projects," *Aircraft Engineering* (September, 1970): 59–66.

6. Bilstein, "Evolution;" David Price, "Political and Economic Factors Relating to European Space Cooperation," *Spaceflight* (January 1962): 7, 10; Walter A. McDougall, "The Scramble for Space," *The Wilson Quarterly* (Autumn 1980): 71–74; European Space Research Organization, *Europe in Space* (Paris: ESRO, 1974), 5–7.

7. Information gathered by author during a series of interviews at ESA headquarters (Paris), European Space Research and Technology Center (Nordwijk, the Netherlands), and ERNO (Bremen, Germany), 26 July–6 August 1982; Herbert Porsche, ed., *10 Jahre Helios* (Bonn: Federal Republic Publishing House, 1984), 11, 159, 164–66.

8. Roger E. Bilstein, "International Aerospace Engineering: NASA Shuttle and European Space Lab," unpublished report, NASA/Johnson Space Center History Office (August 12, 1981), passim; Michael Schwartz, "European Policies on Space Science and Technology, 1960–1978," *Research Policy* 8 (1979): 204–43. The American interest in a European Space Lab is summarized by Wernher von Braun, "Spacelab," *Popular Science* (November 1975): 73.

9. Bilstein, "International Aerospace Engineering," passim; author's interviews at ESA, ESTEC, and ERNO, 26 July–6 August 1982. Also, I gained considerable insight into the European process from an interview with Air Commander A.D.A. Hunley, Society of British Aerospace Constructors, London, United Kingdom (August 10, 1988). For a detailed memoir by the NASA liaison in Europe, see Douglas R. Lord, *Spacelab: An International Success Story* (Washington, D.C.: U.S. Government Printing Office, 1987).

10. Bilstein and Anderson, *Orders of Magnitude*, passim, summarizes principal trends. See also NASA's own detailed annual public relations summary, *Spinoff* (Washington, D.C.: U.S. Government Printing Office), which includes striking photos from various missions; NASA Public Affairs Office, "NASA News/Annual Review," copies in NASA/JSC History Office and NASA Headquarters History Office. My budget figures here are from Aerospace Industries Association, *Aerospace Facts and Figures, 1991–92* (Washington, D.C.: AIA, 1991), 57–59, 63.

11. Bernard C. Nalty, George M. Watson, and Jacob Neufeld, *The Air War over Vietnam: Aircraft of the Southeast Asia Conflict* (New York: Aero, 1981), a book based on reprinted studies by the Office of Air Force History; David A. Anderton, *The History of the U.S. Air Force* (New York: Crescent, 1981), includes an excellent summary, pp. 167–209; Richard P. Hallion, *Storm over Iraq: Air Power and the Gulf War* (Washington, D.C.: Smithsonian Institution Press, 1992), includes a cogent analysis of Vietnam's air war, pp. 17–30.

12. Ray Wagner, *American Combat Planes*, 3rd ed. (Garden City, N.Y.: Doubleday and Company, 1982), 533–36; National Aeronautics and Space Administration, *Progress in Aircraft Design since 1903* (Washington, D.C.: U.S. Government Printing Office, 1974), 77.

13. Wagner, *Combat Planes*, represents a good survey for military designs of the era. Hallion, *Storm over Iraq*, sums up the new operational environment, pp. 35–41, and analyzes the sophisticated elements of "energy maneuverability," pp. 275–81.

14. For a first-rate technical and design summary, see Doug Richardson, *F-16 Fighting Falcon* (New York: Arco, 1983). A knowledgeable editor and journalist covering the European aviation scene, Richardson discusses production and project management, pp. 11–15. For revealing commentary from key individuals on both sides of the Atlantic, I have relied on the transcript of a television documentary, "F-16: Sale of the Century" (1979), produced and written by David Boulton for the series *World*, produced by WGBH Boston for the PBS Network.

15. Robert F. Dorr, "The McDonnell Douglas F/A-18," *World Airpower Journal* I (Spring 1990): 38–73.

16. Jay Steven Ogg, "Metamorphosis of Business Strategies and Air Force Acquisitions Policies in the Aerospace Propulsion Industry: Case Study of the 'Great Engine War'" (master's thesis, Alfred P. Sloan School of Management, 1987); "Dogfight," *New York Times Magazine* (February 1984): 13; Bill Gunston, *World Encyclopedia of Aero Engines* (London: Patrick Stevens, 1989) 69–71; GE Aircraft Engines, "50 Years of Jet Propulsion" (1992), corporate brochure, author's collection.

17. E.E. Bauer, *Boeing in Peace and War* (Enumclaw, Wash.: TABA Publishing, 1991), 195; Wilbur Morrison, *Donald W. Douglas: A Heart with Wings* (Ames: Iowa State University Press, 1991), 231–34.

18. Douglas Ingells, *The McDonnell Douglas Story* (Fallbrook, Calif: Aero Publishers, 1979), 155–58.

19. Ibid., 159–61; John Newhouse, *The Sporty Game: The High-Risk Competitive Business of Making and Selling Commercial Airliners* (New York: Knopf, 1983), 133–37. Morrison, *Donald W. Douglas*, 215, passim.

20. Ronald Miller and David Sawers, *The Technical Development of Modern Aviation* (New York: Praeger, 1970), 203, passim.

21. Laurence S. Kuter, *The Great Gamble: The Boeing 747* (University, Ala.: University Press, 1973), is a concise, informative memoir by a retired Air Force general who had commanded the Military Air Transport Service and joined Pan Am in 1962, becoming vice president, technical staff, in 1965 and retiring again in 1970. Quote from Kuter, p. 11.

22. Bauer, *Boeing*, 278–79; Clive Irving, *Wide Body*, passim. The latter stands out as a superb history of the 747 program, noting the interaction between the manufacturer and its customers.

23. Bauer, *Boeing*, 244–53, 285–89; Newhouse, *Sporty Game*, 169–70, 181; William Green, *Modern Commercial Aircraft* (New York: Random House, 1988), 72–75.

24. Ceruzzi, *Beyond the Limits*, 138–53; Elizabeth A. Muenger, *Searching the Horizon: A History of Ames Research Center, 1940–1976* (Washington, D.C.: U.S. Government Printing Office, 1985), 173–76.

25. On high-bypass ratio engines, see Gunston, *Encyclopedia of Aero Engines*, passim.; Green, *Modern Commercial Aircraft*, includes a useful topical section on high-bypass ratio engines, pp. 32–37. The 757 and 767 are summarized in Bauer, *Boeing*, 309–15, and Green, *Commercial Aircraft*, 80–83.

26. NACA, *Progress*, 80; Francis, *Piper*, 128–33, 185, 188; Weick, *Ground Up*, 328–62; Mark R. Twombly, "Designing for Tomorrow," *AOPA Pilot* 32 (October 1989): 83–84.

27. Mark E. Cook, "Land of Plenty," *AOPA Pilot* 32 (October 1989): 93–95; miscellaneous issues of *Flying: Annual and Buyer's Guide* from the 1970s and early 1980s. The latter journal, issued monthly for many years by Ziff-Davis, reviewed annual trends and catalogued light planes of all descriptions, along with their performance and prices. See, for example, Richard Collins, ed., *Flying: Annual and Buyer's Guide* (1979), passim.

28. Francis, *Piper*, 226–31; Cessna, corporate brochures in author's files; Mark R. Twombly, "M. Stuart Millar," *AOPA Pilot* 32 (October 1989): 215–16.

29. Miscellaneous corporate press releases in author's files. William H. McDaniel, *The History of Beech* (Wichita, Kans.: McCormick-Armstrong, 1982), 309–10, 318–19, 326–28, 357–58, 425; quote from Thomas B. Haines, "Innovation," *AOPA Pilot* 32 (October 1989): 108.

30. Thomas A. Horne, "Dealing with the Feds," *AOPA Pilot* 32 (October 1989): 87–92; Thomas B. Haines, "Innovation," ibid., 107.

31. Cook, "Land of Plenty," ibid., 95.

CHAPTER 8

1. "Internationalization . . . ," *Aerospace Industries Association Newsletter* (July 1988): 1–2.

2. "Aerospace . . . Exporter," Aerospace Industries Association, *AIA Facts and Perspectives* (July 1986), 1–4; Keith Hayward, *International Collaboration in Aerospace* (New York: St. Martin's, 1986), 2–5.

3. David C. Mowery and Nathan Rosenberg, "Commercial Aircraft Industry," in Richard Nelson, ed., *Government and Technical Progress* (New York: Pergamon, 1982), 116; Thomas J. Bacher, "International Collaboration" (paper presented at Conference of Society of Japanese Aerospace Companies, Tokyo, March 29, 1983), 4–5.

4. William B. Scott, "Japanese Industry Shows New Strength, Versatility," *Aviation Week and Space Technology* 116 (May 17, 1982): 163–66. Cited hereafter as *AWST*.

5. National Aeronautics and Space Administration, *Progress in Aircraft Design since 1903* (Washington, D.C.: U.S. Government Printing Office, 1974), 86; Kenneth Munson, *Airliners since 1946* (New York: Macmillan, 1975), 150–51; "Douglas to Build DC-9," *AWST* 78 (April 15, 1963): 40–42.

6. Douglas Ingells, *The McDonnell Douglas Story* (Fallbrook, Calif.: Aero Publishers, 1979), 179–83.

7. Hayward, *International*, 95; "Boeing 767 and the Italian Connection," *Air International* 23 (October 1982): 170, 174–75.

8. Douglas Ingells, *L-1011 Tristar and the Lockheed Story* (Fallbrook, Calif.: Aero Publishers, 1973), 178, 208–13; John Newhouse, *The Sporty Game: The High-Risk Competitive Business of Making and Selling Commercial Airliners* (New York: Knopf, 1983), 173–84; Harold B. Meyer, "The Salvage of the Lockheed L-1011," *Fortune* 83 (June 1971): 66–71; "Down to the Wire with Lockheed," *Fortune* 84 (August 1971): 34.

9. Jack Baranson, *Technology and the Multinationals* (Lexington, Mass.: Heath, 1978), 23–27; Newhouse, *Sporty Game*, 188; "GE's Dream Engine Gets a French Accent," *Business Week* (December 18, 1971): 36; "Stalled Engine Angers the French," *Business Week* (October 21, 1972); 22–24.

10. Baranson, *Technology*, 28–29; Hayward, *International*, 128–37; Virginia Lopez and Loren Yager, *The U.S. Aerospace Industry and the Trend Toward Internationalization* (Washington, D.C.: Aerospace Industries Association, 1988), 40; "Fokker 100," *Air International* 36 (March 1989): 113–15.

11. See, for example, Houston *Chronicle* (September 14, 1984).

12. These examples were culled from miscellaneous press releases and sales brochures from McDonnell Douglas Corporation and Douglas Aircraft Company, 1986–1989, in the author's files. For a popular, recent summary, see Martha Hamilton, "The International Airplane," *Air & Space Smithsonian* 3 (October/November 1988), enclosed as a folded insert with colorful drawings.

13. Bacher, "International Collaboration," 1–4, 7–11; Newhouse, *Sporty Game*, passim; Daniel Todd and Jamie Simpson, *The World Aircraft Industry* (London: Croom Helm, 1986), 56–64; *Policies for Strengthening U.S. Aerospace Trade Performance* (Washington, D.C.: Aerospace Industries Association, 1989), 1.

14. William E. Clayton, Jr., "Fletcher Leaves NASA," Houston *Chronicle* (April 8, 1989); Mark Carreau, "NASA's New Boss," Houston *Chronicle* (April 16, 1989); Craig Couvalt, "NASA Faces Personnel Crisis," *AWST* (February 20, 1989): 23–24; Sylvia Fries, *NASA Engineers and the*

Age of Apollo (Washington, D.C.: U.S. Government Printing Office, 1992), focuses on the 1960s generation, their careers, and reflections of selected individuals as senior managers in the 1980s.

15. Kenneth Gatland, *The Illustrated Encyclopedia of Space Technology* (New York: Harmony Books/Salamander, 1981), 86–95, passim; Frederick I. Ordway III, and Mitchell Sharpe, *Dividends from Space* (New York: Thomas Y. Crowell, 1971); Bill Yenne, *The Encyclopedia of U.S. Spacecraft* (New York: Bison Books, 1985), all provide an informative background.

16. *TRW Space Log: 1989* (Redondo Beach, Calif.: TRW Incorporated, 1990), 15; "Shuttle Flight Set for Next Week," Washington *Post* (November 28, 1992); "Commercial Titan Launch Vehicle Places Two Communications Satellites in Orbit," *AWST* (January 8, 1990): 42–43; "Intelsat F6 Orbited by Commercial Titan," ibid. (July 2, 1990): 25; NASA, "NASA on the Road to Recovery: 1987 In Review," *NASA News* (December 16, 1987).

17. Houston *Post*, "Bush OKs Export" (December 20, 1989); ibid., "Chinese Rocket Carries U.S. Payload" (April 8, 1990); David S.F. Portree, "Thirty Years Together: A Chronology of U.S.–Soviet Space Cooperation" (NASA/Johnson Space Center, 1993, Contractor Report 185707), 30–40; "News Breaks" (*AWST* March 21, 1994): 21; Jeffrey M. Lenorovitz, "U.S. Industry Offers New ELV Proposals," ibid., 24–25.

18. Roger E. Bilstein and Frank W. Anderson, *Orders of Magnitude: A History of NACA and NASA, 1915–1990* (Washington, D.C.: U.S. Government Printing Office, 1989), passim.

19. Howard E. McCurdy, *The Space Station Decision: Incremental Politics and Technological Choice* (Baltimore: Johns Hopkins University Press, 1990), provides a perceptive analysis of that program's evolution. For a comprehensive history of Hubble, see Robert W. Smith, *The Space Telescope: A Study of NASA, Science, Technology, and Politics* (New York: Cambridge University Press, 1989). Kathy Sawyer, "A Closer Look at Eternity," Washington *Post*, National Weekly Edition (April 16–22, 1990): 6–8, captures the flavor of early scientific and public enthusiasm for Hubble. For samples of subsequent disillusions about Hubble, the space station, and NASA in general, see "NASA's Management Woes Severe," Houston *Post* (August 2, 1991); Sharon Begley, "Heaven Can Wait," *Newsweek* (July 9, 1990): 48–55; Dennis Overbye, "Requiem for the Space Station," *Time* (June 10, 1991): 59; Dick Thompson, "Big Gamble in Space," *Time* (February 6, 1993): 62–63. For a programmatic and budget summary, see Bilstein and Anderson, *Orders of Magnitude*, 145–46.

20. Kathy Sawyer, "Space Station Redesign Due by June 1," Washington *Post* (March 11, 1993); William J. Broad, "U.S. to Cut Costs, Seeks Russian Role in Space Station," New York *Times* (April 7, 1993); William B.

Scott, "NASA Aeronautics Budget Fuels High-Speed, Subsonic Research," *AWST* (May 10, 1993): 61.

21. NASA activities compiled from NASA, *NASA: The First 25 Years* (Washington, D.C.: U.S. Government Printing Office, 1983), 27–34; Bilstein and Anderson, *Orders of Magnitude*, 141–50; successive issues of *Spinoff* between 1988 and 1992, as well as the annual summaries of *NASA News*.

22. John Hanrahan, "The Devil and Mr. Jones," *Common Cause Magazine* (November/December 1990): 12–19; "Defense Consultant Guilty in Scandal," Houston *Chronicle* (March 28, 1989); Bauer, *Boeing in Peace and War*, 295–307; Anthony Sampson, *The Arms Bazaar: From Lockheed to Lebanon* (New York: Viking Press, 1977), 222–326; Boulton, *The Grease Machine*, passim. See also N.H. Jacoby, P. Nehemkis, and R. Eells, *Bribery and Extortion in World Business* (New York: Macmillan, 1977).

23. At the core of my commentary in this section about Desert Storm and its technology is a featured report series appearing in the February 1991 issue of *Fortune* magazine. See for example, John Hueg and Nancy Perry, "The Future of Arms," *Fortune* 123 (February 25, 1991): 34–36. A useful, well-illustrated summary of Desert Shield/Storm and its technologies is Time-Life Books, *Air Strike* (Alexandria, Va.: Time-Life Books, 1991); a document that captures public opinion about the war and fascination with its weaponry is "America at War," *Newsweek* (January 28, 1991), a special issue devoted to the conflict, especially the air war section, pp. 15–24.

24. Bill Saporito, "This War Doesn't Mean a Windfall," *Fortune* 123 (February 25, 1991): 40–42; Brenton R. Schlender, "A High-Tech Bird for a Ground War," ibid., 42–42; Bill Gunston, *An Illustrated Guide to Military Helicopters* (New York: Arco, 1981), 64–69.

25. For background and descriptions of hardware and weapons systems developed during the 1980s and eventually used in the Gulf War, see Christopher Chant, *World Encyclopedia of Modern Weapons* (Wellingborough, England: Thorson's Publishing, 1988). See also Michael J. Gething, *Air Power 2000* (London: Arms and Armour Press, 1992), for an informed analysis of Stealth aircraft and other technologies by a professional military journalist. On the evolution of programs and technology leading to systems like the KH-11 surveillance satellites, see William E. Burrows, *Deep Black: Space Espionage and National Security* (New York: Random House, 1986). An informed sampling of Desert Storm's aerospace weaponry and the changing environment of defense contractors appears as a featured issue and cover story, "America's Arsenal," *Fortune* 123 (February 25, 1991): 28–71. Steve Pace, *The F-117A Stealth Fighter* (Blue Ridge Summit, Pa.: 1992), provides a good summary of technical features and operational history. Richard P. Hallion, *Storm over Iraq: Air*

Power and the Gulf War (Washington, D.C.: Smithsonian Institution Press, 1992), represents a perceptive and detailed analysis of the aerial offensive and associated weaponry.

26. John Schwarts, "The $93 Billion Dogfight," *Newsweek* (May 6, 1991): 46–47; Michael A. Dornheim, "Air Force's Hands-Off Approach Speeded ATF Testing Programs," *AWST* (July 1, 1991): 57–58; Gregg Easterbrook, "The Real Lesson of the B-2," *Newsweek* (November 11, 1991): 50–51; Bill Sweetman, "The Stealth Master," *Interavia* 47 (February 1992): 28–33.

CHAPTER 9

1. Statistics and analysis culled from miscellaneous issues of the Aerospace Industries Association *Newsletter* between 1991 and 1993. Consolidation and merger stories appeared frequently in *Aviation Week and Space Technology*; see, for example, the Paris Air Show special issue (June 12, 1989) and the "Annual Forecast and Inventory" issues such as March 18, 1991, and March 14, 1994.

2. Robert Ropelewski, "Rotary Wing Industry in Uncertain Hover," *Interavia* 48 (February 1993): 22–23; Ian Harbison, "Phoenix Patrol," *Helicopter World* 11 (January–March 1992): (reprint); Bruce Smith, "MD Explorer Blends New Technologies," *AWST* 140 (February 22, 1993): "News Breaks," *AWST* 141 (September 12, 1994): 19; corporate releases in author's files.

3. "Aircraft Maker Files Chapter 11," *Houston Post* (July 4, 1991); Edward Phillips, "Piper Calls Chapter 11 an Option," *AWST* (June 3, 1991): 27; Edward H. Phillips, "Stormy Future in Business Flying," *AWST* (September 21, 1992): 40–50; William Gruber, "The Year in Review," *AOPA Pilot* (January 1993): 57–63; "World Roundup," *Aircraft and Aerospace* (April 1993): 31.

4. Anthony Velocci, "Industry Debates Value of Fractional Sales," *AWST* (August 29, 1994): 66–69; Nick Cook, "Prospects Brighten for Special-Purpose Bizjets," *Interavia* 49 (September 1994): 30–34; Robert Ropelewski, "Liability-Reform Bill Brings Relief to General Aviation," Ibid., 28–29; "Hope for General Aviation, *Interavia* 49 (August 1994): 8. There is a cogent summary of the Starship episode in Rowe and Miner, *Borne on the South Wind*, 232–35.

5. "News Breaks," *AWST* (June 26, 1995): 16. In addition to sources cited above, see the special issue of *Interavia* (June 1991) with its cover headline "Connections: Global Links in the Aerospace Industry." Barry Miller's article "Global Links" (pp. 12–25) features numerous charts showing the remarkable spread of the phenomenon. For a summary of the often

convoluted French pattern of overlapping private/state aerospace firms, see Pierre Condom, "A Pivotal Role in Europe," *Interavia* 48 (May 1993): 40–42.

6. Richard G. O'Lone, "777 Revolutionizes Boeing Aircraft Design Process," *AWST* (June 3, 1991): 34–36; Richard Dornheim, "777 Twinjet Will Grow to Replace 747-200," *AWST* (June 3, 1991): 41, 44, 49; Philip Condit, "The 777 Story," *Aerospace* (March 1993): 8–10.

7. Bruce Smith, "Airbus Trims A340 Range Predictions," *AWST* (November 30, 1992): 27–29; Pierre Sparaco, "A321 Rollout Air Airbus Milestone," *AWST* (March 8, 1993): 18–19. On the subsidy issue, see Philip Butterworth Hayes, "When Politicians Become Salesmen," *Interavia* 47 (March 1992): 12–17.

8. "Westinghouse Electronic Systems Wins Russian Air Traffic Control Contract," Washington *Post* (March 27, 1993).

9. John D. Morrocco, "Lockheed Buys Share in Future," *AWST* (December 14/21, 1992): 20–21; Steven Pearlstein, "General Dynamics," Washington *Post* (March 19, 1993).

10. Mark Potts, "Martin Marietta to Acquire Aerospace Division of GE," Washington *Post* (November 24, 1992); Jeffrey Lenorovitz, "GE Aerospace to Merge into Martin Marietta," *AWST* (November 30, 1992): 20–22.

11. David A. Brown, "LTV Aerospace Unit Plans Hard Push for Commercial Subcontract Work," *AWST* (March 2, 1992): 58; "News Summary," *Air and Space* 7 (December 1992/January 1993): 19.

12. "Grumman's Chief Seeks New Course," *AWST* (October 5, 1992): 52–55; James R. Norman, "Ninth Life," *Forbes* (April 26, 1993): 72–74.

13. Robert Ropelewski, "U.S. Industry Continues Downsizing Process," *Interavia* 49 (August 1994): 12–18; Anthony Velocci, "Megamerger Points to Industry's Future," *AWST* 141 (September 5, 1994): 36–38; Robert Samuelson, "Megamergers—Now and Forever," *Newsweek* (September 26, 1944): 51.

14. Aerospace Industries Association, "1993 Year-End Review and Forecast" (press release, December 15, 1993): 1–2; ibid., "1995," (December 1, 1995): 1–3; "Moody's Production Estimates," *Interavia* 49 (August 1994): 8; Paul Proctor, "Boeing Sees New 747, 757 as Vital to Boosting Sales," *AWST* (January 15, 1996): 35.

15. Anthony Velocci, "Douglas Sees Profit in Smaller Volumes," *AWST* 141 (September 19, 1994): 56–59; McDonnell Douglas Corporation, "1993 Annual Report" (St. Louis, Mo., 1993).

16. J.R. Wilson, "Boeing Restructures for Future Uncertainties," *Interavia* 49 (September 1994): 23–26; Ropelewski, "U.S. Industry," 18; Proctor, "Boeing Sees," 35.

17. Philip Parry, "Stacking Up to Compete," *Interavia* 49 (September 1994): 39–40; Michael Mecham, "Douglas Vows to Hold Its Position in

China," *AWST* 141 (August 29, 1994): 51; John Crampton, "Japanese Aerospace: Threat or Partner," *Interavia* 48 (February 1993): 13–16; John Morris, "Japan's Future Role in Asian Aerospace," *AWST*, market supplement (August 29, 1994): S1–S15.

18. Roger Burlingame, *March of the Iron Men: A Social History of Union through Invention* (New York: Grosset & Dunlap, 1938) is still a provocative analysis for European origins of Yankee technology. For later examples, see Darwin Stapleton, "Benjamin Henry Latrobe and the Transfer of Technology," pp. 34–44, and Reese Jenkins, "George Eastman and the Coming of Industrial Research in America," pp. 129–141, in Carroll Pursell, ed., *Technology in America: A History of Individuals and Ideas* (Cambridge: MIT Press, 1990). The themes of lightness and adaptability are explored by Burlingame, as well as Daniel J. Boorstin, *The Americans: The National Experience* (New York: Vintage, 1967). Competition and the systems approach are discussed in Ronald Miller and David Sawers, *The Technical Development of Modern Aviation* (New York: Praeger, 1970), 257–65, 277–79; Bill Gunston, *Bombers of the West* (New York: Scribner's, 1973), 190–92.

19. Bendix is profiled in "William Bendix," *Current Biography* (New York: H. W. Wilson, 1948), 42–44. Robert L. Forward, "Murphy Lives!" *Science 83* (January/February 1983): 78, unveils the real thing. On Gremlin origins, see Roger E. Bilstein, *Flight in America, 317–18,* 378. The Dilbert caricature is recalled in Rosario Rausa, "Jumpin' Jehoshaphat!" *Naval Aviation News* (January/February 1993): 1–9.

20. Robinson is noted in Bill Gunston, *World Encyclopedia of Aircraft Manufacturers* (Annapolis, Md.: Naval Institute Press, 1993), 254, and in "Airscene," *Air International* 44 (March 1993): 118. The sources on Williams include Ken Werrell, *Evolution of the Cruise Missile*, passim, news releases from Williams International, and interviews with curator, Rick Leyes, at the National Air and Space Museum (May 12, 1993).

21. "Aviation, Space, and Defense since 1919," *AIA Newsletter*, 1 (January/February 1989): 9; miscellaneous interviews, author's visit to AIA headquarters (Washington, D.C., 1989); miscellaneous AIA brochures, author's files.

SELECTED BIBLIOGRAPHY

THE FOLLOWING TITLES represent a selected list for quick, convenient reference. Dominick Pisano and Cathleen Lewis, eds., *Air and Space History: An Annotated Bibliography* (New York: Garland, 1988), organizes a startling variety of sources in aeronautics and astronautics that encompasses economic, political, technical, and corporate subjects. Roger Bilstein, *Flight in America: From the Wrights to the Astronauts*, rev. ed. (Baltimore: Johns Hopkins University Press, 1994), considers major trends in both aviation and astronautics and comments on technological patterns.

For the early years of aviation, see Tom Crouch, *The Bishop's Boys: A Life of Wilbur and Orville Wright* (New York: W.W. Norton, 1989), the premier study of the fascinating brothers, their technological achievements, and the formative era of the aviation business. On aspects of air transport development, R.E.G. Davies, *Airlines of the United States since 1914* (Washington, D.C.: Smithsonian Institution Press, 1982), is an indispensable survey with encyclopedic coverage. Welman Shrader, *Fifty Years of Flight: A Chronicle of the Aviation Industry in America, 1903–1953* (Cleveland: Eaton, 1953), provides a chronological summary and annual statistical information that includes general aviation along with other commercial and military aspects. Gene R. Simonson, *The History of the American Aircraft Industry: An Anthology* (Cambridge, Mass.: MIT Press, 1968), touches on the early years and includes major postwar trends.

Regarding aviation manufacturers, John B. Rae, *Climb to Greatness: The American Aircraft Industry, 1920–1960* (Cambridge, Mass.: MIT Press, 1968), covers finances and contributions of the major producers of airliners and military planes. Jacob Vander Meulen, *The Politics of Aircraft: Building an American Military Industry* (Lawrence: University Press of Kansas, 1991), is a detailed analysis of the aviation industry in the 1920s and 1930s, including civil and military. Ronald Miller and David Sawers, *The Technical Development of Modern Aviation* (New York: Praeger, 1970), covers both U.S. and European developments, and explains the essential technologies that gave rise to successful air transportation, from the piston engine era through the impact of jets and swept-back wings. James Hansen, *Engineer in Charge: A History of the Langley Aeronautical Laboratory, 1917–1958* (Washington, D.C.: U.S. Government Printing Office, 1987), offers a fine study of a government research center, set in the context of sweeping changes in aviation engineering. Charles D. Bright, *The Jet Makers: The Aerospace Industry from 1945 to 1972* (Lawrence, Kans.: Regent's Press, 1978), addresses a critical era of technological change. John Newhouse, *The Sporty Game* (New York: Knopf, 1983), presents a detailed narrative of the challenges and pitfalls inherent in designing, building, and marketing modern jet airliners on a global basis. Keith Hayward, *The World Aerospace Industry: Collaboration and Competition* (London: Duckworth, 1994), is a thoughtful, informed summary on the globalization phenomenon. On the general aviation (light plane) industry, see Frank J. Rowe and Craig Miner, *Borne on the Wind: A Century of Kansas Aviation* (Wichita: Wichita Eagle and Beacon Publishing Company, 1994). Although keyed to Kansas, its coverage of Beech, Cessna, and Learjet located within the state provides an instructive picture of this sector of the industry.

For a comprehensive illustrated survey with informative summaries of U.S. military aircraft of the twentieth century, see Ray Wagner, *American Combat Planes*, 3rd ed. (Garden City, N.Y.: Doubleday, 1982). I.B. Holley, *Buying Aircraft: Materiel Procurement for the Army Air Forces* (Washington, D.C.: Dept. of the Army, 1964), is a definitive analysis of World War II manufacturing and technological development, focusing mainly on the airframe industry and including material on power plants. Robert Schlaifer and S.D. Heron, *Development of Aircraft Engines; Development of Aviation Fuels—Two Studies of Relations between Government and Business* (Boston: Harvard Graduate School of Business Administration, 1950), presents a massive study by two leading practitioners in those respective fields. Bill Gunston, *World Encyclopedia of Aero Engines* (London: Patrick Steven, 1989), is an authoritative work with instructive details. Robert F. Coulam, *Illusions of Choice: The F-111 and the Problems of Weapons Acquisition Reform* (Princeton, N.J.: Princeton University Press, 1977), reflects the nature of many contentious development programs of the later Cold War era. Richard Hallion, *Storm over Iraq: Air Power and the Gulf War* (Washington, D.C.: Smithsonian Institution Press, 1992), not only relates aerospace technology to the conflict, but also comments on significant technical trends during the 1970s and 1980s. Ann Markusen et al., *The Rise of the Gunbelt: The Military Remapping of Industrial America* (New York: Oxford University Press, 1991), analyzes geography, economic factors, and politics, with commentary on many aerospace industry centers. Ernest G. Schwiebert, ed., *A History of the U.S. Air Force Ballistic Missiles* (New York: Praeger, 1965), captures the flavor of the early U.S. missile industry. Edgar M. Cortright, ed., *Apollo Expeditions to the Moon* (Washington, D.C.: U.S. Government Printing Office, 1975), does the same for the Apollo space program, with chapters prepared by key participants. The Pulitzer Prize–winning study by Walter McDougal, *The Heavens and the Earth: A Political History of the Space Age* (New York: Basic Books, 1985), provides a stimulating perspective on the political elements of the space industries of the United States, the former Soviet Union, and the European Space Agency.

INDEX

Academy of Model Aeronautics, 11
advertising, 8, 11, 53, 61–62, 140–41, 148, 153–54, 156
Aerojet General, 114, 116, 120–21
Aeronautical Chamber of Commerce of America, 12, 22, 30, 48, 49; name changed to Aircraft Industries Association, 78
Aerospace Corporation, 120
Aerospace Industries Association, 83, 101, 188, 218, 223
Agricultural aircraft, 157
Airbus Industrie, 138, 177, 180, 194, 199–200, 214–15
Air Commerce Act of 1926, 22, 26, 48
airlines, 26–27, 30–31, 33, 36, 42, 44–46, 48, 56–62, 136–45, 219
Air Mail Act of 1925, 22, 30, 32
Air Mail Act of 1934, 36
Aircraft Industries Association, 78, 80; name changed to Aerospace Industries Association, 83
airships, 7. *See also* balloons; dirigibles
American Rocket Society, 111, 114
Anglo-French Purchasing Commission (WW II), 68–69
Apollo-Saturn program. *See* NASA
Arnold, Henry H. (Hap), 2, 6, 67, 84, 118
Asian market, 209, 218–20. *See also* China; Japan
associational activities. *See* corporatist attitudes
ATF (Advanced Tactical Fighter), 205–7
automotive engineering, compared to aviation, 19, 31–32, 50, 74–75, 84
Avco, 81

Aviation (magazine), 30, 33, 38, 73, 140
Aviation Week and Space Technology, 33. *See also Aviation*
AVRO Canada, 127–28, 133

Balbo, Italo, 41, 47
balloons, 7–8, 241n8
Beech (aircraft), 27, 36, 54, 149, 154, 183–85, 211, 213; Model 17, 5, 3; Model 18, 54, 151; Model 35 Bonanza, 151–52, 185; Starship, 183, 212
Beech, Walter, 35, 52
Bell (aircraft), 90; XS-1, 79–80, 115; XP-59, 85; helicopters, 90, 157–59, 158, 169, 210, 222; V-22 tilt-rotor, 200
Bell (missiles/space), 112, 130
Bell, Alexander Graham, 8
Boeing (aircraft), 27, 33–34, 90–91, 137–38, 160, 206, 213–15, 219; B-9, 43, 46, 66–67; B-17, 62, 67, 75; B-29, 67, 75–76; B-47, 92–93, 142, 145; B-52, 93–94, 142, 145; Clipper, 60; F4B-4, 29; Model 40-A, 42; Model 80, 42; Model 247, 41–48, 56–58; Model 299, 43, 67; Model 307, 60–62; Model 707, 83, 141–44, 177, 189; Model 720, 83; Model 727, 144, 180, 189; Model 737, 179–80; Model 747, 144, 178–80, 188–89, 217; Model 757, 180–81, 217; Model 767, 180–81, 189, 191, 217; Model 777, 214; P-26, 29, 61, 66; Stratocruiser, 135, 138–39
Boeing (missiles/space), 121; Minuteman, 121–22, 161
Boeing, William E., 24, 175

Boeing Vertol, 90–91, 200, 222
Brewster XF2A-1 Buffalo, 66
bribes. *See* scandals
budgets. *See* finances

California Institute of Technology (Cal Tech), 39, 63, 112–14. *See also* JPL (Jet Propulsion Laboratory)
Century of Progress International Exposition of 1933–34, 41–42, 47
Cessna (aircraft), 27, 36, 52–54, 149–50; 183–85, 211–12; 170/72 series, 151–54; Airmaster, 53; Citation, 181; T-50, 54
Cessna, Clyde, 34, 52–53
Chance Vought (aircraft), 34, 45. *See also* Vought Corporation
Chanute, Octave, 3
China, 197–98, 219
Chrysler Corporation, 120, 131, 184
Coburn, James (actor), *Our Man Flint*, 156
commonality, 80, 89–90, 144
communications satellites, 195–98
computers, 87–89, 178, 214, 252n9. *See also* electronics
concurrency, 76, 80, 94, 254n26
Confederation of Independent States, 187, 199, 207, 209, 215
Consolidated Aircraft Company, 11, 81; B-24 Liberator, 67, 74–75. *See also* Convair
contract overruns. *See* scandals
contracting. *See* finances
Convair (aircraft), 81; B-36, 81, 91; B-58, 92–93, 95; F-102, 94; F-111, 90, 95. *See also* General Dynamics
Convair (missiles/space), 116, 119; Atlas, 118, 121–22, 128, 197–98
corporatist attitudes, 48; exports and commerce department, 65; national defense, 71, 74; tooling for manufacturers, 95. *See also* Aeronautical Chamber of Commerce (and successors); Air Commerce Act of 1926; military aviation and aerospace; military-industrial complex; NACA; NASA; U.S. Congress
Crouch, Tom, 5
Curtiss (aircraft), 10, 17, 23; Jn-4 Jenny, 10, 13, 18, 27; P-6E Hawk, 29; P-40, 68–69

Curtiss, Glenn, 7–10, 19
Curtiss-Wright Corporation, 20, 34–35, 53, 68–69, 73, 80–82

dealers, 8. *See also* advertising
de Haviland (aircraft), DH-4, 18–19, 34
de Havilland, Comet jet airliners, 141, 143; D.H. 88 Comet, 58–59
de Seversky, Alexander P., 66, 72
dirigibles, 53, 60
Disney, Walt, 72, 125
diversification, 96
Douglas (aircraft), 25, 56–63, 138, 160, 175–76; A-4 Skyhawk, 97; C-47, 135; C-54, 135; DC-1, 57; DC-2, 56–60; DC-3, 56, 58–60, 62, 138; DC-4E, 61–62; DC-4, 62, 135, 138; DC-6, 135, 139, 144; DC-7, 136, 139, 142; DC-8, 83, 143, 145, 176–77; DC-9, 145, 166, 176, 179–80, 189–90; DC-10, 166, 177, 180, 189–91; merger with McDonnell, 175–77
Douglas (missiles/space), 120–21, 130–32
Douglas, Donald, 10, 25, 175–76

EDO Corporation, 33
education, aeronautical and aerospace engineering, 10, 13–14, 22, 37–39, 51–52
Eisenhower, Dwight D., 101, 109, 119–20, 125, 152
electronics, 87–89, 91, 94, 116–17, 123–24, 152, 161, 169, 180–81, 196, 215
engines (gas turbine), 81–82, 84–87, 141–44, 174–75, 191–93, 206–7, 222–23
engines (piston), 1, 11, 15, 18, 21, 22, 39–40, 33, 40–41, 44–45, 47, 54, 64, 74, 77, 81, 100, 242n15, 243n28
ESA (European Space Agency), 166–68, 185, 198–200. *See also* European aviation/space programs
European aviation/space programs, 90, 127, 137–38, 140, 143–44, 160, 163–68, 172–73, 180–81. *See also* Airbus Industrie; multinational production
European influences: before WW II, 3–5, 10, 12, 14, 16, 19–20, 22–24, 37–40, 42, 47–48, 50, 55–56, 64, 66–67; experience of WW II, 66–70; following WW II, 81–82, 84–87, 91, 93, 99–101, 106, 110–11, 117–18, 125,

127–28, 160, 163–68, 183, 190–91, 212–13, 222
exports: before WW II, 10, 16–17, 22, 24, 29–30, 49, 53–54, 58–59, 65; experience of WW II, 66–70; following WW II, 80–83, 100–101, 106, 112–13, 138, 143–44, 147, 157–59, 170–75, 182–83, 186, 191–95, 210, 215, 218–19, 222–23

federal role in aviation and aerospace. *See* corporatist attitudes
finances: before WW II, 3–4, 6, 13, 16–18, 24, 31–33, 35–36, 42, 44–46, 51–54, 57, 64–66; experience of WW II, 67–70, 75–77; following WW II, 80, 92–93, 101–2, 106, 116, 122, 128–29, 138, 140, 143–45, 147–48, 155, 161, 163–64, 168, 171–74, 178, 182–83, 192–93, 196, 199, 203–5, 207–8, 209–10, 212, 216–18; Wrights and patent dispute, 15–16
float planes, 33
flush riveting, 61, 78
flying boats, 10, 33, 60
Fokker (aircraft), 56, 140, 143
Ford: Tri-Motor, 24, 26; Willow Run plant, 75; aerospace, 105
Fortune (magazine), 52
fuel research, 40–41

GALCIT (Guggenheim Aeronautical Laboratory of the California Institute of Technology), 111–12, 114
Gardner, Lester, 30, 33
Gates Learjet, 156–57. *See also* Learjet
GE (General Electric), 85, 87, 100, 106, 116, 118, 129, 174–75, 192–94, 196, 204, 216
general aviation: before WW II, 8–9, 22, 27–28, 35–36, 50–55; experience of WW II, 52, 54–55; following WW II, 82–83, 145–59, 160–61, 181–86, 211–12, 222–23; liability issues, 184–85, 212. *See also* names of individual manufacturers
General Aviation Manufacturers Association, 136
General Dynamics (aircraft), 81, 103, 106, 206; F-111, 101–4, 172; F-16, 160, 172–75, 187; Lockheed merger, 216

General Dynamics (missiles/space), 197–98. *See also* Convair (missiles/space)
General Motors, 66, 71
geographic location, 36
Gluhareff, Michael, 85–86
Goddard, Robert, 110–11
Gremlins, 222
Grumman (aircraft), 25, 101; and mergers of 1990s, 217; F3F-2, 29; F4F-2 Wildcat, 65–66; F6F Hellcat, 66; F-14, 103, 171–72, 217; Gulfstream, 154
Grumman (missiles/space), 129–30
Grumman, Leroy (Roy), 25, 34, 65
Guggenheim Foundation for the Promotion of Aeronautics, 7, 38–39, 245n17, 248n7
Gulfstream, 184, 211. *See also* Grumman Gulfstream
Gunston, Bill, 95, 221

Haley, Andrew, 114, 116, 255n5
Harker, Ronald W., 70, 100
Hartzell Corporation, 32–33
Heinemann, Edward, 97
helicopters, 33, 90–91, 136, 157–59, 169, 203–5, 210–11, 222–23
Heron, Samuel D., 40
Holley, I. B., 75
Hoover, Herbert, 48
Hughes Helicopter, 90
Hunsaker, Jerome, 9, 14, 25, 38, 49, 112

IBM (International Business Machines), 89, 114
ICBM (intercontinental ballistic missile). *See* missiles (military)
IGY (International Geophysical Year), 108, 125
infrastructure, 11, 18, 22, 28–30, 32–33, 48, 113
Institute of Aeronautical Sciences, 49–50
international collaboration, 163–68, 187–88, 198–200, 207, 222–23. *See also* exports; IGY; multinational production; START
IRBM (intermediate range ballistic missile). *See* missiles (military)
Irving Parachute Company, 33

Jacobs, Eastman, 63, 69
Jamouneau, Walter C., 51

Japan, 53, 60, 100, 189–90, 198, 211, 214, 219–20
JATO (Jet Assisted Takeoff), 112, 114–15
Jet aircraft: airliners 136–37, 139–45; corporate jets, 136, 155–59, 181; origins, 84–87. *See also* engines (gas turbine); military aviation following WW II; names of individual manufacturers
jets. *See* engines (gas turbine); jet aircraft
Johnson, Clarence (Kelly), 26, 97–99
Jones, Robert, 86
JPL (Jet Propulsion Laboratory), 112–13

Kaman Aerospace, 90–91, 210, 222
Kaman, Charles, 90
Kartveli, Alexander, 66–67
Kennedy, John F., 109, 126–27
Klemin, Alexander, 38

labor, 35–36, 71–74, 82, 84, 218
lean production, 223
Lear, Bill, 136, 155, 157
Learjet (aircraft), 155–59; merger with Bombardier, 211. *See also* Gates Learjet
light plane industry, 22, 27, 34, 50–55, 146. *See also* general aviation
Lindbergh, Charles, 21, 40–41; "Lindbergh boom," 26, 30
Link, Edwin, 33
Lockheed (aircraft), 25–26, 55–56, 59, 61, 70, 106; and mergers of 1990s, 216–17; C-5A, 102, 178; Constellation, 61–62, 94, 135–36, 138–39, 144; Electra (turboprop), 83, 140; F-22, 206–7; F-104, 89, 97–101, 166, 172; F-117, 205, 207; P-38, 97; P-80, 85, 98; Tristar, 180, 191–92. *See also* Johnson, Clarence (Kelly); Skunk Works
Lockheed (missiles/space), 123
Lockheed brothers, (Allan and Malcolm), 25
Loening (aircraft), 25, 34, 55–56
Loening, Grover, 6, 11–13, 15–16, 19, 25, 30, 34

MacRobertson England to Australia Air Race, 58–59
McDonnell Douglas (aircraft), 106, 203–6, 218–19; AH-64 Apache, 203–5, 210, 214–15, 218; AV-8A Harrier, 172; F-4 Phantom II, 169–71, 177; F-15 Eagle, 171, 189; F-18, 174; MD-11, 194, 214; MD-80 series, 219; MD-90 series, 219; MD-520 series; 210–11
McDonnell Douglas (missiles, space), 117, 130–31, 196–97
McDonnell Douglas, corporate merger, 176–77
McDonnell James, 176–77
McNamara, Robert, 101–3
McNary-Watres Act, 42–43
Malina, Frank, 111–14
management, 4–48, 73, 75–76, 97–101, 120–21, 123–24, 127–30, 132–33, 174–75, 203–5. *See also* concurrency; lean production; Packard Commission; teaming
marketing. *See* advertising; airlines; public attitude
Martin, Glenn, 9–10, 13, 25
Martin (aircraft), 139–40, 141; B-10, 43, 66–67; M-130, 56, 60
Martin (missiles/space). *See* Martin Marietta
Martin Marietta, 118, 162, 164, 196–98; mergers of 1990s, 216–17; Peacekeeper, 162; Titan, 116, 121, 128, 196–97
Maule (aircraft), 185, 222
mergers, 11, 34, 81, 117, 149, 160, 177, 184–85; key mergers of the 1990s, 211, 215–18
military aviation and aerospace: before WW II, 3–6, 7–8, 10–20, 24–28, 30–31, 43, 46, 59, 63–67; experience of WW II, 65–78, 137–38; following WW II, 79–107, 109, 112–24, 160–65, 168–75, 201–8, 213; Gulf War, 187, 202–5; Vietnam, 168–71. *See also* missiles (military)
military-industrial complex, 80–81, 101
minorities, in aircraft industry, 72–73
missiles (military), 83, 88–89, 104–7, 113–24, 160–65; cruise missiles, 162; missile gap, 109. *See also* finances; NASA; space exploration; submarine/missile systems
MIT (Massachusetts Institute of Technology), 9–10, 25, 38, 123
Mooney, 154, 184–85
MTOW (maximum takeoff weight), 55–56, 220
multinational production, 100–101, 160, 166–67, 172–75, 188–95, 211, 213, 219–20, 222–23. *See also* exports; IGY
Murphy's Law, 221–22

NACA (National Advisory Committee for Aeronautics), 7, 14–15, 20, 22, 29–30, 37–39, 43, 48, 63–65, 68–69, 78, 79, 83, 86–87; cowling, 37–38, 43, 57, 63; reorganization as NASA, 83–84
NASA (National Aeronautics and Space Administration), 163–68, 195–200; Apollo-Saturn program, 116, 126–34, 163; origins, 83–84, 98, 125–26; space shuttle, 164–67, 196; space station, 98, 168, 198–99
National Defense Education Act, 109
New York World's Fair of 1939, 66
North American (aircraft), 73–74; and merger with Rockwell, 117; F-86, 92, 189; Navion, 149; P-51 Mustang, 68–70, 74; X-15, 115
North American (missiles/space), 116–17, 119, 121, 123, 132
Northrop (aircraft), 25, 101, 172, 174, 201, 205–6; mergers of 1990s, 217
Northrop (missiles/space), 119, 123; influence on computers, 87–89
Northrop, John (Jack), 25, 40, 43

O-ring, 55

Packard Commission, 203, 205–6
Parallel development, 120–21
Pawlowski, Felix W., 38
Peacekeeper (MX missile), 121, 161–62
Piper (aircraft), 50–52, 81, 149, 154, 181–83, 211; Cherokee, 181; Coupe, 52; Cruiser, 52, 150; J-2 Cub, 50–51; Pacer, 150, 152; Pawnee, 157
Piper, William T., 51
Polaris. *See* submarine/missile systems
Pratt, Fletcher, 111
Pratt & Whitney, 34, 40, 44–45, 84–85, 87, 93, 174–75, 192–94, 206–7, 211
pressurization, 61
profits. *See* finances
Project Paperclip, 86–87, 91–92
propellers, 1, 22, 32–34, 41
public attitude, 2–3, 5, 9, 19, 21–22, 41–42, 46–48, 52, 67, 71–72, 108–9, 125–27, 134, 220–22

Rachmaninoff, Sergei, 33

RAND Corporation, 79
Raytheon 105, 183–84, 205, 211–12
Reaction Motors Incorporated, 114–16
Reconstruction Finance Corporation, 71
Republic Aviation (aircraft), 66–67; P-47, 66–67, 148; Seabee, 148–49
retractable gear, 25, 29, 34, 43, 45, 55
roadable airplanes, 149
Robinson Helicopters, 222–23
Rockefeller, Laurance, 115
Rockefeller, William, 50
Rocketdyne, 116, 117, 120, 130
rocketry. *See* missiles; space exploration
Rockwell (aircraft): B-1, 174–75, 193. *See also* North American; Rocketdyne
Rohrbach, Adolph, 39
Rolls-Royce, 69–70, 87, 100, 191–92, 223
Roosevelt, Franklin D., 67, 71
Royal Aeronautical Society, 50
Royal Air Force, 70, 77, 106, 133, 170

satellites. *See* communications satellites; Sputnik; spy satellites
scandals, 18–19, 100–101, 200–201, 203, 205
Schmued, Ed, 69
Schneider Trophy Races, 22–24
Schriever, Bernard S., 119–20
Seversky (aircraft), 66–67; P-35, 66. *See also* Republic Aviation
shared-risk production, 145
Sidewinder, 104–7, 171
Sikorsky (aircraft), 33, 90; helicopters, 90, 159, 210
Sikorsky, Igor, 33, 44, 90
simulators, 33, 89
Skunk Works, 97–98
Smith, Richard K., 220
Snow, Leland, 157
social/cultural aspects: Art Deco styling, 59. *See also* advertising; Century of Progress International Exposition; Disney, Walt; Gremlins; Murphy's Law; public attitudes
space exploration, 83, 108–34, 160, 163–65, 166–67
space station, 98, 168
Sputnik, 108, 133, 166
spy satellites, 197, 205

START (Strategic Arms Reduction Talks), 163, 185
Stearman (aircraft), 27, 34, 36, 157
Stearman, Lloyd, 52
Stinson, 53, 81, 149
Strategic Defense Initiative (SDI), 162–63, 185
stressed-skin construction, 39–40, 43
strikes. *See* labor
subcontractors, 18, 32–33, 45, 75, 96, 116, 122, 188, 217
submarine/missile systems, 122–24, 132–33, 163
systems management, 120, 221. *See also* Polaris; weapon systems

teaming (development/production), 173–74, 206–7, 213
Thiokol, 105, 115–16, 121, 125, 164
Thomas, B. Douglas, 10
Thomas brothers (Oliver and William), 10–11, 24
Thomas-Morse Aircraft Corporation *See* Thomas brothers
tooling, 45–46, 95–96, 98–99, 142–43
Trippe, Juan, 50, 142
TRW (Thompson-Ramo-Wooldridge), 120–21

unions. *See* labor
United Aircraft Corporation, 33–34, 43, 64
U.S. Air Mail Service, 26
U.S. Congress, and aircraft industry, 19, 34, 76–77, 97. *See also* corporatist attitudes

vendors. *See* subcontractors; infrastructure
Victory through Air Power (de Seversky),72
Volta Congress, 85
von Braun, Wernher, 87, 117–18, 123, 125, 127, 132–33
von Karman, Theodore, 39, 63, 111–12, 114, 118
von Neumann, John, 119
von Ohain, Hans, 84, 86
Vought Corporation, 216–17. *See also* Chance Vought
V-2 rocket, 116–17

Waco (aircraft), 28, 53
Wagner, Herbert, 39
Wallace, Dwane, 52–53
Warner, E. P., 37, 48–49, 61
weapon systems, 75–76, 94. *See also* systems management
Whittle, Frank, 84–85, 86
Williams International, 222–23
Willow Run, 75
Wings Club, 50, 83
women, in aircraft industry, 35, 72–73. *See also* labor
workers. *See* labor
Wright, Theodore P., 73, 137–38
Wright Aeronautical Corporation, 40. *See also* Curtiss-Wright Corporation
Wright brothers (Orville and Wilbur), 1–7, 9, 12, 19, 32
Wright, Katherine, 5

About the Author

Dr. Roger Bilstein is Professor of History Emeritus at the University of Houston-Clear Lake, a suburban, upper-division campus for juniors, seniors, and graduate students. He has written numerous articles and entries in encyclopedias and reference works, as well as eight books that cover many aspects of the aviation and aerospace industries and space exploration. His honors include appointments to the Charles Lindbergh Chair of Aerospace History at the Smithsonian Institution's National Air and Space Museum and as a visiting professor of history at the U.S. Air Force's Air War College. Dr. Bilstein has also served as a consultant and guest curator for several museums in the United States. Since his retirement from the classroom, he continues to write and to work as a consultant at his home in the Texas Hill Country near Austin, Texas.